# So You Think You Know

# So You Think You Know

## Secrets of the Numbers

Benjamin Jacob Bloch, Ph.D.

Volume III

Library of Congress Control Number:


ISBN: 978-157780036
1517780039

*In the beginning was space*
*and space was with anti-space*
*and space was anti-space*
*as in the beginning*

## Mystical Math

☆ ♀ ≈ Φ ✡ Φ

# About the Author

Benjamin Jacob Bloch holds a Ph.D. in Theoretical Physics from The Polytechnic Institute of Brooklyn; a Professional Degree in Education; Permanent Certification 9–12 in Math, Physics and Science; and Certification in Supervisor/Administrator in New York State. A retired Adjunct Full Professor of Physics at the Polytechnic Institute of New York, he is the founding Advisor to Omega Phi Alpha. At Commack High School he taught AP Physics and AP Mathematics, mentored teachers and developed new programs. A former concert violinist he is a graduate of the High School of Music & Art. His industrial experience includes Columbia University Hudson Laboratories, the Bendix Corporation, General Precision, the Grumman Aerospace Corporation and a scientific mission to the North Polar Region. For eight years he published a monthly math problem and contributed editorials to the Professional Surveyor Magazine.

Books by this author:
All A's All Ways, Marduk Publishing Inc. 2005
So You Think You Know Series, Amazon :Publishing
       Volume 1 DaVinci and Kabbalah, 2015,
       Volume II The Divine Duality, 2016
       Volume III Secrets of the Numbers, 2023

Dedicated to my students

# CONTENTS

## Volume I

# Volume II

# Volume III

# Definitions

The Real Numbers

≡ means defined as

Natural ≡ The counting numbers {1, 2, 3, …}

Whole ≡ Natural numbers including 0

Integer ≡ Positive and negative whole numbers {…, -3, -2, -1, 0, 1, 2, 3, …}

Rational ≡ Numbers that can be expressed as a ratio of an integer to a non-zero integer. All integers are rational, but the converse is not true.

Irrational numbers ≡ Real numbers that are not rational


SDQ  Single Digit Quality ≡ The quality of a number reduced to a single digit

Notation for SDQ is as follows. 41=> 5

AQ Addition Quality

RB  Repeat Block ≡ Repeating decimal digits

P  Period ≡ Number of digits in a Repeat Block

RBS Repeat Block Sum ≡ Sum of the Repeat Block digits

SDQ(RBS) ≡ The SDQ of the Repeat Block Sum indicated by **bold italic**.


N.B. Digits that form a periodic pattern are indicated by a bold italic.

Double Prime ≡ A prime number that is also an SDQ prime.

Triple Prime ≡ A prime number having SDQ of two more primes

Example 59,999 => 41 => 5


Sequences

Fibonacci: $a_n = a_{n-1} + a_{n-2}$   $a_1 = 1, a_2 = 1$

Lucas: $a_n = a_{n-1} + a_{n-2}$   $a_1 = 1, a_2 = 3$

Padovian: $a_n = a_{n-2} + a_{n-3}$   $a_0 = a_1 = a_2 = 1$

Pell: $a_n = 2a_{n-1} + a_{n-2}$   $a_0 = 0, a_1 = 1$

Pell-Lucas: $a_n = 2a_{n-1} + a_{n-2}$   $a_0 = a_1 = 2$

Perrin: $a_n = a_{n-2} + a_{n-3}$   $a_0 = 3, a_1 = 0, a_2 = 2$


Basic Principles

All numbers represent both quantity and qualities

Number qualities include: biology, chemistry, physics, and music associations

Even numbers have female characteristic, odd numbers male

The smaller the number quality the greater its potency

The lost Hebrew gematria has been recovered. (See Volumes 1 and 2.)

Currently there are three kinds of numbers: cardinal, ordinal and nominative.

Cardinal numbers are the numbers that are used for counting. It helps us to know how many elements are there.

Ordinal numbers are the concept of natural numbers which is used to describe a way to arrange different elements.

Nominal numbers can be defined as numbers that are used for identifications.

We will explore an expanded role for number called the Quality of number.

Indeed number qualities are evident in many applications. For example, in chemistry the Periodic Table identifies the characteristics of elements by number. The number of protons in a nucleus, the atomic number is a number quality. The number of elements in an atomic shell determines the potency of that element, the fewer the number of electrons the greater the potency.

[FAR] For Additional Research

Some follow up research is designed for the readership. For example in the section on inverse numbers in calculating the Repeat Block Sum.

# Chapter 1

## Background

Our study goes back thousands of years prior to the ancient Greeks, into the fourth millennium before the common era, to the Sumerians whose origin is unknown and to the Semites who continued in their works. Their systems of mathematics; the sexigesimal system, logarithms, exponentials, irrational numbers and music tuning theory were already in advance of all but present day knowledge. A thousand years before Abraham and well before the Hebrew Scriptures they enjoyed roles for number as elements characterizing both physical and nonphysical systems. Representing a physical system it had the quality of a specific force or energy; in a nonphysical system it had the quality of a principle, or state of being. Lost over time, number quality is a significant component of ancient knowledge.

Pythagoras, Plato, Newton, and Einstein, are among the more recent scholars seeking to rediscover the elemental building blocks forming the physical and nonphysical worlds. They investigated the basic principles governing the relationships between number, color, music, and form.

Presented here is a system to advance this fundamental quest. Three principles govern any prospective system; simplicity, versatility and consistency. Simplicity in concept and utilization, versatility in scope and application and consistency throughout. The building blocks developed here must conform to these three principles.

The diverse aspects of number developed here are key to unlocking the universal connections binding sound and sight to spirit and form. Missing from all prior works, spirit, is the essential ingredient of both the nonphysical and physical worlds. Indeed, spirit awakened here within number finds in music its ultimate expression. Number alone are the elemental building blocks of the universe. For the ancients, the single digits 0 though 9 are the "Single Deity Qualities" of the universe, herein referred to as Single Digit Quality (SDQ).

The goals for this work are; to identify the many roles of number, to explore them, and thereby reveal their beauty as the only elements requisite for the formation and maintenance of all forms. Finally, the roles of number, as related to music, provide the missing connections between the physical and nonphysical worlds. Number, are thus archetypes, spanning many dimensions of meanings, and yielding their particular flowers according to their applied fields of context. This work begins by recovering the ancient roles of number, examining their relation to mathematics, music, physics, and finally to magic squares. Further research into these and additional areas of study are given as suggested research.

## Number Duality

Numbers are currently used to count objects or things. Adding more objects increases the amount, or quantity of objects. Acting in this capacity, number represents a quantity. This limited role for number, taken for granted and unquestioningly accepted as complete, was not always so. Number in its ancient applications had numerous additional meanings: as related to the stars, as Deities, as sexual qualities; as elements of music theory, as elements of physics, and ultimately as the essential building blocks of the spiritual and physical universe. Today, only that single aspect of number, namely to numerate objects, has survived the crucibles of time.

### *NUMBER: QUANTITY and QUALITY*

We define two broad categories of number; quantity; and quality. For example, the number 19,624; has a quantity of 19,624; potencies of 19,624 and (1+9+6+2+4) 22, and (2+2) 4; and a quality of 4. We are already familiar with the role of number as quantities. As qualities; number could represent sexual principles, physical principles, physical properties, or musical qualities. Number can also have the quality of form, related to the sides of a figure, or points, or aspects of shape. For every quality, number can also indicate the degree of intensity, or potency. We will examine in detail those lost aspects of number; quality and potency.

Chemistry inherently utilizes number: as a quantity, representing the total number of electrons; as a quality, representing the number of electrons outside a closed shell, and therefore related to the property of the atom; and as a potency, describing chemical activity.

As quality, all elements having the same number of electrons in the outer shell of the atoms that form that element, are said to belong to the same family of elements.

As potency, each member of the same chemical family, has a different total number of electrons, the smaller the total number, the greater its chemical activity, or potency. For example, elements belonging to the chemical family known as the Halogens, have the same number of electrons in their outermost shell, and the potency of each is relative to the total number of electrons in the atom. The smaller the total number of electrons, the greater the activity of the element. Thus, total number acts as a potency. The role for the number zero in chemistry, is a quantity for elements having zero electrons in their outer shell. These elements, such as the Nobel gases have a completed electronic shell, and belong to the same family. The potency of the inactive Nobel gases, in turn depends upon the total number of electrons in the atom, the fewer that total, the greater its relative potency.

Number to the ancients was related to the gods, and could also represent a physical quality; as mass, speed, power and so on. The smaller the number, the greater its physics quality. Zero, is then the most powerful physics quality, followed by one, etc. The applications of the roles of number to the Mesopotamian gods is reserved for future publication.

The extended role for number as both quantity and quality is an important feature of a Divine Science. As a quantity, number is completely objective representing the number of times a single quantity is repeated. Thus, as a quantity, 10 signifies 1 taken 10 times. In this usual role, it is purely objective and abstract. However as a quality, number signifies both an aspect and an essence, usually expressed as a type of personality or behavior. One can say therefore, that number specifies and spans both body and spirit.

The extensive role of number satisfies a dimension of physics missing at present. For example Newton's Law of Gravitation, or Coulomb's law, are inverse square laws, wherein the force between two masses or two charges is inversely proportional to the square of the distance separating the masses or charges. The mathematical relationship produces a number for the level of the force but does not answer the question of why that force should exist in the first place. Physics is used to try and answer how, but not why. Physics can describe a force and how it varies, but it cannot be used to explain why force should exist.

The field created by a mass or charge permeates all space, the force that the field is capable of exerting is never exactly zero, no matter how great the finite distance from the mass or charge. To all practical purposes, since the force approaches zero as the distance increases, that force is considered negligible at large distances. In quantum mechanics, however, the wave function that characterizes the system, and which is used to calculate probabilities, is never really zero. Typically in an experiment such as the scattering of particle by an atom, or another particle, a judgement is made is to when the wave functions can be considered negligible. For practical reasons one attempts to solve the simplest approximation to a real system first. Then if successful, additional constraints are added. Science is therefore in part subjective, a condition rarely if ever examined or discussed.

It is precisely the subjective aspect of science, quantum mechanics, that the quality of number describes. For example, if a mass is moving, in other words has momentum, it has a wavelength, known as the deBroglie wavelength, ($\lambda$ = h/mv, where h = 6 x 10-34 joules x sec, is Max Planck's constant, m is mass, and v is speed). The larger the momentum, the smaller its corresponding deBroglie wavelength, and the smaller the momentum, the larger the deBroglie wavelength.

Thus, if the momentum is extremely small, the corresponding wavelength is very large, and can span a great distance. As the momentum approaches zero, the wavelength becomes infinite. Thus if a raindrop on a leaf in the Amazon forest is disturbed, its deBroglie wavelength could extend to alpha Centauri. The actual magnitude of this disturbance can be quite negligible, whatever the mode of comparison, but it is not zero. It is the quality of this disturbance that is addressed by the quality of number.

Meditation centers around silencing the activities of the body and the mind, thus reducing related momenta, and thereby extending any associated wavelengths. It is possibly the quality aspect of these wavelengths that allow communication with and within the universe. Let us examine some of the different aspects of the quality number, starting with sexual qualities.

## Number: Sexual Qualities

As we have noted, in the remote past, number also had a quality of sex. In its widest range of physical and nonphysical application, the ancients considered number quality not only in terms of physical sexual properties, but also as sexual principles. A sexual quality of number was associated with it properties of being even or odd. An even number was associated with the female and an odd number, with the male. It must be emphasized that the range of sexual qualities denoted by female and male, was not restricted in either scope or application.

Female, as a sexual quality of number could mean receptivity, or passivity, or balance or a host of other meanings, depending upon the particular application. Male, as a sexual quality of number could mean activity, or generative faculties, or fertilization, or many other meanings, also depending upon the particular application.

**Number: Potency**

The potency aspect of number is determined by continued reduction to a single digit. The smaller the number quality, the greater its potency. This is different from number in its role of quantity, and where it acts as a counter. Thus, the most potent number, according to this definition, is the female number, zero. The next most potent number, then, is the male number, one. A little reflection reveals that this applies to the Periodic Table of Elements in modern chemistry. Classification of elements with similar chemical features is by families, each family containing the same number of outer electrons. The number of electrons in the outermost shell of an atom within and element, determines the location of that element within various families of elements. All member elements of a particular family, contain the same quantity of electrons in their outer shells. Within each family, the smaller the total number of electrons in the atom, the more active, or more potent, that element. Thus the periodic table represents elements both in terms of number quantity and number quantity.

**Number: Biological Quality**

How many sexes are there? According to Anne Fausto-Sterling, geneticist and professor of medical science at Brown University, investigators recognize the concept of inter sexual bodies, so that biologically there are many sexes between the male and the female extremes. There are true male-female combinations, and dominant male and dominant female with secondary female and male characteristics, both internal and external. Plato recognized three sexes; male, female, and hermaphrodite. Hermaphrodite, from: herms = one testis and one ovary (the sperm and egg producing vessels or gonads). There are additional characteristics of hermaphrodites, namely:

Merms ≡ male pseudo-hermaphrodite some testes & aspect of female genitalia, but no ovaries.

Ferms ≡ female pseudo-hermaphrodites have ovaries and some aspects of the male genitalia, but lack testes.

The sexual qualities of numbers are easily determined by their even and odd properties. Even numbers are female, and odd numbers are male. Thus 1 is male, 2 is female, 3 is male, until and including 9 being male. We employ the SDQ to determine all of the sexual characteristics of numbers. Now the number 10 being even is female, but it also has an SDQ of 1, ( may be reduced to 1 by adding digits), which is odd and thus male. Consequently, 10 has a primary female quality, and a secondary male quality. This is similar to the dominant and recessive properties in genetics. We shall denote these qualities by writing the sexual characteristics of the 10 as Fm, primary Female and secondary male. Then, 11 is Mf, 12 is Fm, until we get to 19. Now 19 is Male, 9+1=10 is female, and 1+0=1 is male. Thus 19 is written as Mfm, a dominant male, a secondary female and also a tertiary male characteristic.

Example: Consider the number 219:
As a quantity, it numerates 219 objects or things.
As a potency quality, it equals 219, and 2+1+9 = 12, and 1+2 =3. It therefore has three potencies; 219, 12, and 3.
It has the single digit quality of 3.
It has the sexual qualities of; 219 odd thus primary Male, 12 even thus secondary female, and 3 odd thus tertiary male. We therefore write its sexual qualities as Male female male, or simply Mfm.

It is clear that we can apply aspects of quality to any number derived from any mathematical operation. Using the number 219 as a further example:
$\sum 219 = 0+1+2+...+219 = 24,090$, which has the potencies of 24,090, 15, and 6, and the single digit quality of 6. It therefore has the biological quality of Fmf.

Aspect Versus Essence
There is an ancient saying, "Three in aspect, One in Essence." In biology, a member of a species looks a certain way, depending upon that member's genetic makeup. However, a member's looks, only reveal the dominant aspect of its genetic makeup, and thus there are the terms genotype and phenotype. Genotype refers to the actual biological makeup, while phenotype refers to the dominant aspect of that makeup that distinguishes that member. For example, a member might have two dominant genes, or a dominant and a recessive gene.

The individual characteristic controlled by that gene, will appear the same in both cases, the phenotype, yet its biological makeup will not be the same, the genotype. The aspect is then like the phenotype, and the essence, like the genotype.

So too for numbers. A number may have a potency and a different quality. The potencies are associated with the aspect, and relate to the number itself. The quality is associated with the Single Digit Quality, which is the essence of that number. Thus, number plays a greater role than has been utilized in the modern era. The size of a number, or its quantity, or phenotype, is a potency, as is the number when reduced to two or more digits. When the number is finally reduced to its single digit, the SDQ, it now shows its essence, its genotype. This is most easily seen by a direct reference to the sexual aspects of number, with its phenotype sex, its quantity, or potency, and its secondary characteristics, its genotype, essence, or SDQ.

The potency aspect of any number is determined by its magnitude, the smaller the number, the greater its potency. This is the opposite of a number in its role purely as a quantity, where it serves merely as a counter. Therefore, the most potent number is the female (even) number, zero. This association of the potent female zero is related to the power of "mother nature," where zero might represent; the womb, or the earth, or the cell. The quality of zero can also be analyzed in terms of its shape; a finite inside, a circumference, and a semi infinite outside - as the ancients said, "Three in aspect, one in essence."
The next most potent number, then, is the male number, one. A little reflection reveals that this applies to the Periodic Table of Elements in modern chemistry.

Classification of elements with similar chemical features is by families, each family containing the same number of outer electrons. The number of electrons in the outermost shell determines the position of that element within the families of elements. All member elements of a particular family contain the same quantity of electrons in their outer shells. Elements with fewer electrons are more active, or potent, than elements with more electrons. Thus, the periodic table represents elements both in terms of number quantity and number potency.

# Chapter 2

## *Single Digit Quality (SDQ)*

We have seen that number, have characteristics to specify chemical properties of elements within the periodic table. To find the family quality of a number, simply reduce the number to a single digit. Thus, the number 103 has a quantity of 103, and the quality of 103 and 4. We will use the symbol => to indicate the reduction of a number to its digits, thus, 15,546 => 21 => 3.

The cycle of number single digit quality is 9, (modulo 9), since the number 10 reduces to a quality of 1, (10 = 1 + 0 = 1). The number quality of 11, is 2, of 12, 3, and so forth. Thus all number qualities lend themselves to arrangements similar to those of the periodic table of elements. Note that the number 0, has a unique number quality, and will be discussed as a special case. We can form the following table of number quality, having a cycle of 9:

| 0 | | | | | | | | |
|---|---|---|---|---|---|---|---|---|
| 1 | 2 | 3 | 4 | 5 | 6 | 7 | 8 | 9 |
| 10 | 11 | 12 | 13 | 14 | 15 | 16 | 17 | 18 |
| 19 | 20 | 21 | 22 | 23 | 24 | 25 | 26 | 27 |

and so forth...

Thus 1, 10, 19, all have the number quality of 1. Each column has the same number quality. Drawing the analogy to chemistry, each column, with its identical number quality, belongs to the same family. We note that within each family, or quality, as we move down that column, the number potency increases. The smaller the actual number, the greater the potency. Thus in the first column, of number quality 1, 1 is more potent then 10, 10 is more potent then 19, and so forth. Again, 1, 10, and 19, all have the same number quality, namely, 1. This method of number analysis is called Single Digit Quality, or simply, SDQ. We will come back to examine significance of the single digit quality in more detail, later on.

Number: Properties Of Single Digit Quality

The "single digit quality," or SDQ system consists of adding the digits of any number, until a single digit is reached. This single digit is a quality, and not just a quantity. Thus, the number 6,130, becomes 6+1+3+0, which becomes 10, which then becomes 1+0, which finally becomes 1. Therefore, the SDQ, or single digit quality of the number 6,130, is 1. We will write this as: SDQ(6130) = 1, or 6,130 => 1. In all of number, there are only ten possible single digit qualities, corresponding to the ten digits of 0 to 9. Let us examine the properties of SDQ in detail. Table 1 is arranged according to families of equal number quality. Note that 0 is unique, since no other number has a SDQ of 0. The numbers in each of the nine columns have the quality of the single digit at the top of the column. Thus, the number whose quantity is 41, has the quality of 5, while the number whose quantity is 128, has the quality of 2.

TABLE 2-1 NUMBERS AS QUALITIES

| 0 | | | | | | | | |
|---|---|---|---|---|---|---|---|---|
| 1 | 2 | 3 | 4 | 5 | 6 | 7 | 8 | 9 |
|  |  |  |  |  |  |  |  |  |
| 10 | 11 | 12 | 13 | 14 | 15 | 16 | 17 | 18 |
| 19 | 20 | 21 | 22 | 23 | 24 | 25 | 26 | 27 |
| 28 | 29 | 30 | 31 | 32 | 33 | 34 | 35 | 36 |
| 37 | 38 | 39 | 40 | 41 | 42 | 43 | 44 | 45 |
| 46 | 47 | 48 | 49 | 50 | 51 | 52 | 53 | 54 |
| 55 | 56 | 57 | 58 | 59 | 60 | 61 | 62 | 63 |
| 64 | 65 | 66 | 67 | 68 | 69 | 70 | 71 | 72 |
| 73 | 74 | 75 | 76 | 77 | 78 | 79 | 80 | 81 |
| 82 | 83 | 84 | 85 | 86 | 87 | 88 | 89 | 90 |
| 91 | 92 | 93 | 94 | 95 | 96 | 97 | 98 | 99 |
|  |  |  |  |  |  |  |  |  |
| 100 | 101 | 102 | 103 | 104 | 105 | 106 | 107 | 108 |
| 109 | 110 | 111 | 112 | 113 | 114 | 115 | 116 | 117 |
| 118 | 119 | 120 | 121 | 122 | 123 | 124 | 125 | 126 |
| 127 | 128 | 129 | 130 | 131 | 132 |  |  |  |

Let us examine features of this table of numbers arranged according to SDQ. Other than 0, there are nine families. The numbers shown in non bold script, reduce directly to a single digit, while those in the bold script, require more than one step in the SDQ reduction.

Example:10 = 1; 20 = 2; 42 = 6, and so forth, while 37 => 10 > 1; 55 => 10 => 1; 58 => 13 => 4.

Any number has a quality of being either odd or even. If it is odd, it is considered to represent a male quality, if even, a female quality. Furthermore, any number, by simply adding its digits, is reducible to a smaller number. If that reduction produces an even number, than the secondary sexual quality is female, is odd than male. The greater the number of reductions, the greater the number of sexual qualities. For example, consider the number 126. It is even and thus has a primary female characteristic. When reduced, 1+2+6 = 9, it becomes odd, which denotes male characteristics. Thus the number 126 is considered as Fm, firstly female and then male, dominantly female and recessively male.

The numbers in bold script shown in the table, can be multiply reduced, thus 65 => 11 => 2, and thus 65 is Male, male, female (65 = odd, male, 11 = odd, male, and 2 = female), thus finally, 65 = Mmf.

The unreduced or raw number sexual property is shown capitalized. As another example, the number 1999 => 19 => 10 => 1, and is thus Mmfm. Thus all numbers exhibit primary, secondary, and tertiary, etc., sexual characteristics. By the simple techniques of digit reduction, together with even and odd properties, the sexual characteristics of all numbers can be determined.

Other than 0, there are nine quality families. The numbers shown in lightface type reduce directly to a single digit, while those in the boldface type require more than one step in the SDQ reduction.

## DOUBLE DIGIT QUALITIES

There are nine single digit qualities, that all numbers possess, together with the zero, which is unique in that no other number reduces to it, and therefore it stands alone. There are ninety double digit numbers that reduce to the nine single digit qualities, numbers 10 through and including 99.

If we examine the double digit numbers that have the same quality as 1, we see that these may be categorized as: four symmetric number pairs, (a symmetric pair are two numbers whose digits are in reverse order, 19 and 91); the number 10; and finally the number 55, which is formed by doubling the number 5. Again, these are (19, 91), (28, 82), (37, 73), (46, 64), and 10 and 55. Each number, of course, has the SDQ of 1.

Looking at the double digit numbers that have the SDQ of 5, we get the 50 and the double digit 77, and the symmetry pairs: (14, 41), (23, 32), (59, 95), and (68, 86). [rem: examine if there are patterns and how that pattern repeats for triple digit numbers.]

## SDQ OF PRODUCTS

Table 2-2 is the usual multiplication table.

TABLE 2-2 MULTIPLICATION TABLE

|    | 1  | 2  | 3  | 4  | 5  | 6  | 7  | 8  | 9   | 10  | 11  | 12  |
|----|----|----|----|----|----|----|----|----|-----|-----|-----|-----|
| 1  | 1  | 2  | 3  | 4  | 5  | 6  | 7  | 8  | 9   | 10  | 11  | 12  |
| 2  | 2  | 4  | 6  | 8  | 10 | 12 | 14 | 16 | 18  | 20  | 22  | 24  |
| 3  | 3  | 6  | 9  | 12 | 15 | 18 | 21 | 24 | 27  | 30  | 33  | 36  |
| 4  | 4  | 8  | 12 | 16 | 20 | 24 | 28 | 32 | 36  | 40  | 44  | 48  |
| 5  | 5  | 10 | 15 | 20 | 25 | 30 | 35 | 40 | 45  | 50  | 55  | 60  |
| 6  | 6  | 12 | 18 | 24 | 30 | 36 | 42 | 48 | 54  | 60  | 66  | 72  |
| 7  | 7  | 14 | 21 | 28 | 35 | 42 | 49 | 56 | 63  | 70  | 77  | 84  |
| 8  | 8  | 16 | 24 | 32 | 40 | 48 | 56 | 64 | 72  | 80  | 88  | 96  |
| 9  | 9  | 18 | 27 | 36 | 45 | 54 | 63 | 72 | 81  | 90  | 99  | 108 |
| 10 | 10 | 20 | 30 | 40 | 50 | 60 | 70 | 80 | 90  | 100 | 110 | 120 |
| 11 | 11 | 22 | 33 | 44 | 55 | 66 | 77 | 88 | 99  | 110 | 121 | 132 |
| 12 | 12 | 24 | 36 | 48 | 60 | 72 | 84 | 96 | 108 | 120 | 132 | 144 |

Table 2-3 is an SDQ multiplication table, since all of the products are reduced to their SDQ numbers. Remember that to get the SDQ of any number, simply add that number's digits until a single digit is obtained. Thus, SDQ(36) = 9; or using another notation, 36 => 9.

TABLE 2-3 SDQ MULTIPLICATION TABLE

|    | 1 | 2 | 3 | 4 | 5 | 6 | 7 | 8 | 9 | 10 | 11 | 12 |
|----|---|---|---|---|---|---|---|---|---|----|----|----|
| 1  | 1 | 2 | 3 | 4 | 5 | 6 | 7 | 8 | 9 | 1  | 2  | 3  |
| 2  | 2 | 4 | 6 | 8 | 1 | 3 | 5 | 7 | 9 | 2  | 4  | 6  |
| 3  | 3 | 6 | 9 | 3 | 6 | 9 | 3 | 6 | 9 | 3  | 6  | 9  |
| 4  | 4 | 8 | 3 | 7 | 2 | 6 | 1 | 5 | 9 | 4  | 8  | 3  |
| 5  | 5 | 1 | 6 | 2 | 7 | 3 | 8 | 4 | 9 | 5  | 1  | 6  |
| 6  | 6 | 3 | 9 | 6 | 3 | 9 | 6 | 3 | 9 | 6  | 3  | 9  |
| 7  | 7 | 5 | 3 | 1 | 8 | 6 | 4 | 2 | 9 | 7  | 5  | 3  |
| 8  | 8 | 7 | 6 | 5 | 4 | 3 | 2 | 1 | 9 | 8  | 7  | 6  |
| 9  | 9 | 9 | 9 | 9 | 9 | 9 | 9 | 9 | 9 | 9  | 9  | 9  |
| 10 | 1 | 2 | 3 | 4 | 5 | 6 | 7 | 8 | 9 | 1  | 2  | 3  |
| 11 | 2 | 4 | 6 | 8 | 1 | 3 | 5 | 7 | 9 | 2  | 4  | 6  |
| 12 | 3 | 6 | 9 | 3 | 6 | 9 | 3 | 6 | 9 | 3  | 6  | 9  |

## FEATURES OF THE SDQ PRODUCTS

It is immediately noticed that there is a pattern to each horizontal row and each vertical column. In fact, even the 10, 11, 12, etc, may be written as their SDQ numbers, see the Minimum SDQ Multiplication Table. Thus, only a 1 through 9 table is required to produce the SDQ values for the product, and for every set of factors. [23]

A. 1x pattern:

As expected the 1x pattern is just 1,2,3,4,5,6,7,8,9; looking across the first row. This is a nine term pattern.

B. 2x pattern:

The 2x pattern is 2,4,6,8,1,3,5,7,9; looking across the second row. This is a nine term pattern.

C. 3x pattern:

The 3x pattern is 3,6,9; a three term pattern.

D. 4x pattern:

The 4x pattern is 4,8,3,7,2,6,1,5,9: a nine term pattern.

E. 5x pattern:

The 5x pattern is 5,1,6,2,7,3,8,4,9; a nine term pattern.

Other than the 9's this 5x pattern is the reverse of the 4x pattern.

F. 6x pattern:

The 6x pattern is 6,3,9; a three term pattern.

G. 7x pattern:

The 7x pattern is 7,5,3,1,8,6,4,2,9; a nine term pattern.

Other than the 9's this 7x pattern is the reverse of the 2x pattern.

H. 8x pattern:

The 8x pattern is 8,7,6,5,4,3,2,1,9; a nine term pattern.

Other than the 9's this 8x pattern is the reverse of the 1x pattern.

I. 9x pattern:

The 9x pattern is all 9's.

EXAMPLE, consider the product of 4,631 x 5,872 = 27,193,232. Taking SDQ of each factor, we get SDQ(4,631) = 5, and the SDQ(5,872) = 4. Looking at the table of SDQ products, we see that SDQ(5) x SDQ(4) = SDQ(2). Thus the SDQ(27,193,232) should equal SDQ(2), which it does. (SDQ(27,193,232) = SDQ(2+7+1+9+3+2+3+2) = SDQ(29) = SDQ(2)).

TABLE 2-4 MINIMUM SDQ MULTIPLICATION TABLE

|   | 1 | 2 | 3 | 4 | 5 | 6 | 7 | 8 | 9 |
|---|---|---|---|---|---|---|---|---|---|
| 1 | 1 | 2 | 3 | 4 | 5 | 6 | 7 | 8 | 9 |
| 2 | 2 | 4 | 6 | 8 | 1 | 3 | 5 | 7 | 9 |
| 3 | 3 | 6 | 9 | 3 | 6 | 9 | 3 | 6 | 9 |
| 4 | 4 | 8 | 3 | 7 | 2 | 6 | 1 | 5 | 9 |
| 5 | 5 | 1 | 6 | 2 | 7 | 3 | 8 | 4 | 9 |
| 6 | 6 | 3 | 9 | 6 | 3 | 9 | 6 | 3 | 9 |
| 7 | 7 | 5 | 3 | 1 | 8 | 6 | 4 | 2 | 9 |
| 8 | 8 | 7 | 6 | 5 | 4 | 3 | 2 | 1 | 9 |
| 9 | 9 | 9 | 9 | 9 | 9 | 9 | 9 | 9 | 9 |

Notice that by shading the all of the 3, 6 and 9's each two by two set of boxes has an SDQ of 4 when multiplied on the diagonal and all of the numbers in each box add to an SDQ of 9..

For example, the first box is

| 1 | 2 |
|---|---|
| 2 | 4 |

And we see that: 1x4=4 and 2x2=4; and 1+2+2+4 =9.

The next box to the right is

**4        5**

**8        1**

And here we note that 4x1=4, and 5x8=40=>4. Additionally, we note that 4+5+8+1 =18=>9.

All the other boxes have the same SDQ properties.

## TEST FOR INCORRECTNESS

Thus, the results of any multiplication may be instantly tested for incorrectness, by reducing each multiplier to its SDQ and then using the table to determine the SDQ of its product. If they do not agree, then the multiplication is in error. There is no limit to the size of each multiplier. Furthermore, interchanging any digits in either multiplier does not effect the SDQ outcome of the product. There are many other features of this table that relate to prime numbers that will be discussed in the next chapter.

# Chapter 3

## SDQ DIVISION

Consider the case where c=b/a. In this form SDQ is undefined, in other words, we **cannot** write SDQ(c) = SDQ (SDQ(a)/SDQ(b)). However, if we write this in the mathematically equivalent form ac=b, then SDQ for multiplication can be used in certain cases to check on the **incorrectness** of the division result.

Since SDQ(SDQ(a) x SDQ(c)) = SDQ(b), then for the operation of division, where c=b/a this should apply.
Example: If b=8 and a=2, then c=b/a=8/2=4. Writing this example as a product b=ac or 8 = 2 x 4. In this form SDQ(8) = SDQ(SDQ(2) x SDQ(4)).

We ask the question: what two number products results in an SDQ of 1?
Examining the minimum SDQ multiplication table we note that:
Products that yield an SDQ of 1:
1x1=1, 2x5=1, and 4x7=1 are the only products that will yield 1.
Writing these in division format we get 1/1=1, 1/5=2, ½=5, ¼=7 and 1/7=4.
83x41=3403, or 2x5=1.

Any cell contents when divided by its row contents equals its column contents, and any cell contents when divided by its column contents, equals its row contents. Simply locate any SDQ value in the table, and note its two SDQ row and column values. For example, if we look at the contents in row 5 and column 6, we find a 3. The SDQ 3, when divided by its row SDQ 5, equals its column SDQ 6; and similarly, the SDQ 3, when divided by its column SDQ 6, equals its row SDQ 5.

**GENERAL FEATURES OF SDQ DIVISION**
Table 3-1 shows the results of SDQ division. Please note that all numbers are SDQ values.
A. 1/1=1; 1/2=5; 1/3 missing; 1/4=7; 1/5=2; 1/6 missing; 1/7=4; 1/8=8; 1/9 missing.
B. 2/1=2; 2/2=1; 2/3 missing; 2/4=5; 2/5=4; 2/6 missing; 2/7=8; 2/8=7; 2/9 missing.
C. 3/1=3; 3/2=6; 3/3=1,4,7; 3/4=3; 3/5=6; 3/6=2,5,8; 3/7=3; 3/8=6; 3/9 missing.
D. 4/1=4; 4/2=2; 4/3 missing; 4/4=1; 4/5=8; 4/6 missing; 4/7=7; 4/8=5; 4/9 missing.
E. 5/1=5; 5/2=7; 5/3 missing; 5/4=8; 5/5=1; 5/6= missing; 5/7=2; 5/8=4; 5/9 missing.
F. 6/1=6; 6/2=3; 6/3=2,5,8; 6/4=6; 6/5=3; 6/6=1,4,7; 6/7=6; 6/8=3; 6/9 missing.

G. 7/1=7; 7/2=8; 7/3 missing; 7/4=4; 7/5=5; 7/6 missing; 7/7=1; 7/8=2; 7/9 missing.

H. 8/1=8; 8/2=4; 8/3 missing; 8/4=2; 8/5=7; 8/6 missing; 8/7=5; 8/8=1; 8/9 missing.

I. 9/1=9; 9/2=9; 9/3=3,6,9; 9/4=9; 9/5=9; 9/6=3,6,9; 9/7=9; 9/8=9; 9/9=1,2,3,4,5,6,7,8,9.

TABLE 3-1 SDQ DIVISION

| 1 | 1 | 2 | 3 | 4 | 5 | 6 | 7 | 8 | 9 |
|---|---|---|---|---|---|---|---|---|---|
| 2 | 5 | 1 | 6 | 2 | 7 | 3 | 8 | 4 | 9 |
| 3 |   |   | 1,4,7 |   |   | 2,5,8 |   |   | 3,6,9 |
| 4 | 7 | 5 | 3 | 1 | 8 | 6 | 4 | 2 | 9 |
| 5 | 2 | 4 | 6 | 8 | 1 | 3 | 5 | 7 | 9 |
| 6 |   |   | 2,5,8 |   |   | 1,4,7 |   |   | 3,6,9 |
| 7 | 4 | 8 | 3 | 7 | 2 | 6 | 1 | 5 | 9 |
| 8 | 8 | 7 | 6 | 5 | 4 | 3 | 2 | 1 | 9 |
| 9 |   |   |   |   |   |   |   |   | ALL |

We immediately see that we cannot divide by 3, 6 or 9 without either getting no result or multiple results. Division by all other digits is single-valued. Later we shall show that any number greater than 3, having an SDQ of 3, 6, or 9 is not a prime number.

## SDQ OF FRACTIONS

The ancient Egyptians regarded the division of 1 as the basis for all fractions. We now examine the division of 1 relative to Repeat Block RB, Periods P and Single Digit Quality, SDQ. For example, 1/7 = 0.**142857**142857 is a repeating fraction with the Repeat Block **RB** = **142857** with digits **1 4 2 8 5** and **7** and decimal repeat block period **P** = 6 digit digits. The digits 3, 6 and 9 are not present. To find the SDQ of the digits in this repeat block, simply continue to add the digits of the resulting sum until a single digit obtains. For the example given: 1+4+2+8+5+7= **27=> 9**, so that the **SDQ** for this **RBS** sum equals **9**. We shall employ the SDQ symbol => to mean the reduction to a single digit. In this study of fractions we seek any patterns that emerge.

## Definitions:

RB  Repeat Block ≡ The decimal digits in a Repeating Fraction  in **bold italic**

P ≡ Period or number of digits in a Repeat Block.

RBS ≡ Repeat Block Sum, the sum of the digits in a Repeat Block

SDQ(RB) ≡ The SDQ of the Repeat Block Sum in **bold italic**

N.B. Prime number denominators ≥ 7 are in **bold**.


**SDQ and Inverse Primes**

Statement: No Repeat Block Period is greater than the denominator minus 1.

P ≤ D-1 [independent of the numerator].

[DO ANY OF THE REPEAT BLOCKS of 2 or more digits CONTAIN THE DENOMINATOR?] Use underline to indicate!

**Verify: Division of 1 by a prime number equal to or greater than 7 produces a repeat block whose SDQ is always 9.**

This does not mean that division of 1 by a non prime number cannot also yield an SDQ of 9. Such as 14, 21, 22, 26, 28, 34, 35, 38, 39. 42, 44, 46, 49, 52, 55, 56, 57, 58, 62, 65, 68, 70, …

Multiples of 7 are in italics and all yield SDQ 9, except for 7 times multiples of 9. Also 2 times any prime also yields a SDQ 9.

**Decimal Types**

**A. Non repeating digits**

      1) Finite number of digits, such as 1/8=0.125

      2) Infinite number of digits, such as e, π and all irrational numbers, √2 = 1.414…etc.

**B. Single repeating digit, such as all 3's, as in 1/3=0.333333333333**

The possible single repeating digits are all: 1, 2, 3, 5, 6, or 7.

      All 1's: 1/9, 1/90, 1/900, …

      All 2's: 1/45,

      All 3's: 1/3, 1/12, 1/30, 1/48, 1/75, 1/120, … SDQ of each denominator, 3

      All 4's: **NONE**

      All 5's: 1/18,

      All 6's: 1/6, 1/15, 1/24, 1/60, 1/96, … SDQ of each denominator, 6

      All 7's: 1/36,

      All 8's: **NONE**


**C. Repeat Block Sum**

      1) Repeat Block sum whose single digit quality, SDQ is: 1, 3, 6, or 7

          SDQ = 1: 1/27, 1/81, …

          SDQ = 3: 1/33, 1/102, 1/123, 1/126, …

          SDQ = 6: 1/51, 1/63, 1/66, 1/69, 1/87, 1/117, …

          SDQ = 7: 1/108,

      2) Repeat Block Sum whose single digit quality, SDQ is 9. **Prime denominators are shown in bold type.**

**1/7**, **1/11**, 1/12, 1/14, **1/17**, 1/19, 1/21, 1/22, **1/23**, 1/26, 1/28, **1/29**, **1/31**, 1/34, 1/35, **1/37**, 1/38, 1/39, **1/41**, 1/42, **1/43**, 1/44, 1/46, **1/47**, 1/49, 1/52, **1/53**, 1/55, 1/56, 1/57, 1/58, **1/59**, **1/61**, 1/62, 1/65, **1/67**, 1/68, 1/70, **1/71**, 1/73, 1/74, 1/76, 1/77, 1/78, **1/79**, 1/82, **1/83**, 1/84, 1/85, 1/86, 1/88, **1/89**, 1/91, 1/92, 1/93, 1/94, 1/95, **1/97**, 1/98, **1/101**, **1/103**, 1/104, 1/105, 1/106, **1/107**, **1/109**, 1/110, 1/111, 1/112, **1/113**, 1/114, 1/115, 1/116, 1/118, 1/119, 1/121, 1/122, 1/124, **1/127**,

**Every prime number ≥ 7 has an inverse whose RBS (Repeat Block) SDQ is 9.**

### Another Pattern

Consider the number 3.14159. The first digit is 3. If we add the next digit (1) to 3 we get 4. Adding the next digit (4) to 4 we get 8. Adding the next digit (1) to 8 we get 9. Adding the next digit (5) to 9 we get 14 with a SDQ of 5. Adding the next digit (9) to 5 we get 14 with a SDQ of 5. In this fashion we collect all of these digits to yield, 3.48955. Essentially we are adding all of the preceding digits, always taking their SDQ values and forming a new number.

### First Few Prime Numbers

**2**  **3**  **5**  **7**  **11**  13  17  19  **23**  **29**  31  37  **41**  **43**  **47**  53  **59**  **61**  67  71  73
**79**  **83**  89  **97**  **101**  103  107  109  **113**  127  **131**  **137**  139  **149**  **151**  157
163  **167**  **173**  179  181  **191**  193  197  19  211  **223**  **227**  229  233  **239**  **241**
251  **257**  **263**  269  271  **277**  **281** 283  **293**  307  **311**  **313**  **317**  **331**  337  **347**
**349**  **353**  359  **367**  373  379  **383**  **389**  397  **401**  409  **419**  **421**  431  433  **439**
**443**  449  **457**  461  463  467  **479**  487  **491**  499  503  **509**  521  523  541

Doubly primes are also **SDQ prime** and shown here in **bold italic** type.

### The Doubly Prime Numbers

2, 3, 5, 7, 11, 23, 29, 41, 43, 47, 59, 61, 79, 83, 97, 101, 113, 131, 137, 149, 151, 167, 173, 191, 223, 239, 241, 257, 263, 277, 281, 293, 311, 317, 331, 347, 349, 353, 367, 383, 389, 401, 419, 421, 439, 443, 457, 461, 479, 491, 509,

**There are 15 doubly primes less than 100; 24 less than 200; 32 less than 300; 41 less than 400; 50 less than 500.**

**SDQ Values for the doubly prime numbers are:**
2, **3**, 5, 7, 2, 5, 2, 5, 7, 2, 5, 7, 7, 2, 7, 2, 5, 5, 2, 5, 7, 5, 2,

**REMARKABLE RESULTS TO BE VERIFIED AS A GENERAL RULE**
It appears that other than the number 3, the only SDQ values for doubly prime are 2, 5, and 7.


Instant test of primes....
[FAR] Perform the usual tests on primes and then do on their SDQ primes. Then apply SDQ 3 rule.
**Analyze all similar repeat blocks, in particular, P 6.**
**Do all P = 6 have an SDQ of 9? No, for example: 1/63, 1/117, 1/126. All these seem to have the same denominator SDQ of 9.**
**N.B. Analyze the relationship between the denominator and 1 less than the P. When does it hold and is it only for the doubly prime?**
**In other words, When is P = D – 1, where D = denominator?**


**P = D – 1**
Inverses with above property
7, 17, 19, **23**, **29**, **47**, **59**, 61, 97, 109, 113, 131, 149, 167, 179, 181, 223, 229, 233, 257, 269, 313, 337, 347, 373,
499, 503, 509, **not 521**, 523, 541,
1019,    7901, 7919,


[FAR] P = D - 1      Do all these RB Sum => 9?
All denominators having this property end in 1, 3, 7 or 9.
P=>D check pattern


**First Few Prime Numbers (Bold type is used for primes that are also doubly prime)**

| | | | | | | | | | | | | | | | |
|---|---|---|---|---|---|---|---|---|---|---|---|---|---|---|---|
| 2 | 3 | 5 | 7 | **11** | 13 | 17 | 19 | **23** | **29** | 31 | 37 | **41** | **43** | **47** | 53 | **59** |
| **61** | 67 | 71 | 73 | **79** | **83** | 89 | **97** | **101** | 103 | 107 | 109 | **113** | 127 | **131** | **137** |
| 139 | **149** | **151** | 157 | 163 | **167** | **173** | 179 | 181 | **191** | 193 | 197 | 199 | 211 | **223** |
| 227 | 229 | 233 | **239** | **241** | 251 | **257** | **263** | 269 | 271 | **277** | **281** | 283 | **293** | 307 |
| **311** | **313** | **317** | **331** | 337 | **347** | **349** | 353 | 359 | **367** | 373 | 379 | **383** | **389** | 397 |
| **401** | 409 | **419** | **421** | 431 | 433 | **439** | **443** | 449 | **457** | **461** | 463 | 467 | **479** | 487 |
| **491** | 499 | 503 | **509** | 521 | 523 | 541 |

Question: Do any non prime denominators have this property of **P = D - 1** where **D** is the denominator?

Question: Do any inverses have an **SDQ(RB) = SDQ(D)**? Or written as **[P=>D]**

We begin with the examination of the division of 1.
The following lists the Fraction, its decimal equivalent, period P and the Single Digit Quality of the repeat block sum, SDQ(RB). Prime denominators are in **bold** and doubly primes in **bold italic**.

| FRACTION | P | SDQ(RB) |
|---|---|---|
| 1/1 = 1 | 1 | |
| 1/**2** = 0.5 | 5 | |
| **1/3** = 0.333333… | all 3's | |
| 1/4 = 0.25 | 7 | |
| **1/5** = 0.2 | 2 | |
| 1/6 = 0.1666666… | all 6's | |
| | | |
| **1/7** = 0.142857142857 | P 6 | 27=>9 [P=D-1] |
| 1/8 = 0.125 | 8 | |
| 1/9 = 0.111111 | | all 1's |
| 1/10 = 0.1 | 1 | |
| **1/11** = 0.090909 | P 2 | 09=>9 [P=>(D)] |
| 1/12 = 0.08333333… | all 3's | |
| **1/13** = 0.076923076923 | P 6 | 27=>9 |
| 1/14 = 0.0714285714285 | P 6 | 27=>9 |
| 1/15 = 0.0666666... | all 6's | |
| 1/16 = 0.0625 | 13=>4 | |
| **1/17** =0.05882352941176470588235294117647 | P 16 | 72=>9 [P=D-1] |
| 1/18 = 0.0555555... | all 5's | |
| **1/19** = 0.052631578947368421052 | P 18 | 81=>9 [P=D-1] |
| 1/20= 0.05 | 5 | |
| 1/21= 0.047619047619 | P 6 | 27=>9 |
| 1/22 =0.0454545454545… | P 2 | 45=>9 |
| **1/23**=0.043478260869565217391304 34 | P 22 | 99=>18=>9 [P=D-1] |
| 1/24=0.04166666666666… | all 6's | |
| 1/25=0.04 | 4 | |
| 1/26=0.0384615384615… | P 6 | 27=>9 |
| 1/27=0.037037037037037… | P 3 | 10=>1 |
| 1/28=0.03571428571428 | P 6 | 27=>9 |
| **1/29**=0.0344827586206896551724137931 03448 | P 28 | 126=>9 [P=D-1] |

1/30=0.033333333333333…                          all 3's
**1/31**=0.**0322580645161290**32258064516129032      P 15    **54=>9**
1/32=0.03125                                     11=>2
1/33=0.030303030303030303030303030303030303      P 2     3
1/34=0.0**2941176470588235**2941176470588235       P 16    **72=>9 [P=>D]**
1/35=0.0**285714**2857142857142857142857142**9**     P **6**     **27=>9**
1/36=0.02777777777777…                           all 7's
**1/37**=0.0**270270270270270270270270270270270**27    P 3     **27=>9**
1/38=0.0**26315789473684210**526315789473684      P 18    **81=>9**
1/39=0.0**25641025641025641025641025641**026       P **6**     **18=>9**
1/40=0.025                                        7
**1/41**=0.0**2439**02439024390243902439024390**24**     P 5     **18=>9 [P=>D]**
1/42=0.0**238095238095238095238095238095**24       P **6**     **27=>9 [P=>D]**
**1/43**=0.0**2325581395348837209302325581395**3      P 21    **90=>9**
1/44=0.0**22727272727272727272727272727272**72      P 2     **9**
1/45=0.0**22222222222222222222222222222222**22      all 2's
1/46=0.0**2173913043478260869565**2173913043      P 22    **99=>18=>9**
**1/47**=0.0**2127659574468085106382978723404255319148936170**2127659574468
                                                 P 46    **207=>9  [P=D-1]**
1/48=0.0208333333333333333333333333333333333      all 3's
1/49=0.0**20408163265306122448979591836734693877551**020408163265306122…
                                                 P 42    **189=>18=>9**
1/50=0.02                                         2
1/51=0.0**1960784313725490**196078431372549       P 16    **69=>15=>6**
1/52=0.01**9230769230769230769230769230769**       P **6**     **27=>9**
**1/53**=0.0**1886792452830**1886792452830188679    P 13    **63=>9**
1/54=0.0**185185185185185185185185185185**18       P 3     14=>5
1/55=0.0**1818181818181818181818181818181**8       P 2     **18=>9**
1/56=0.017**857142**8571428571428571428571**43**      P **6**     **27=>9**
1/57=0.0**1754385964912280**7017543859649122       P 18    **81=>9**
1/58=0.0**172413793103448275862068965517**24       P 28    **126=>9**
**1/59**=0.0**169491525423728813559322033898305084745762711864406779661**0169491
                                                 P 58    **261=>9 [P=D-1]**
1/60=0.0166666666666666666666666666666666       all 6's
**1/61**=0.0**1639344262295081967213114754098360655737704918032786885245**901639
34426…                                           P 60    **270=>9 [P=D-1]**
1/62=0.0**161290322580645**16129032258064516       P 15    **54=>9**
1/63=0.0**15873**015873015873015873015873015       P **6**     24=>6

| | | |
|---|---|---|
| 1/64=0.015625 | | 19=>10=>1 |
| 1/65=0.0**153846**153846153846153846153846615 | P **6** | **27=>9** |
| 1/66=0.0151515151515151515151515151515 | P 2 | 6 |
| **1/67**=0.0**149253731343283582089552238805970**149 | P 32 | **144=>9** |
| | | |
| 1/68=0.0**1470588235294117647**0588235294118 | P 16 | **72=>9** |
| 1/69=0.0**144927536231884057971**01449275362 | P 22 | 96=>15=>6 |
| | | |
| 1/70=0.0**142857**142857142857142857142857 14 | P **6** | **27=>9** |
| **1/71**=0.0**1408450704225352112676056338028169**0140 | P 35 | **126=>9 [P=>D]** |
| 1/72=0.013888888888888888888888888888888 | | all 8's |
| **1/73**=0.0**13698630**1369863013698630136863 | P 8 | **36=>9** |
| 1/74=0.0**135**1351351351351351351351351351 3513 | P 3 | 9 |
| 1/75=0.013333333333333333333333333333333 | | all 3's |
| 1/76=0.0**13157894736842105263**1578947 36842 | P 18 | **81=>9** |
| 1/77=0.0**12987**0129870129870129870129 87013 | P **6** | **27=>9** |
| 1/78=0.0**128205**128205128205128205128205 13 | P **6** | **18=>9 [P=>D]** |
| **1/79**=0.0**1265822784810**1265822784810126582 | P 13 | **54=>9** |
| 1/80=0.0125 | 8 | |
| 1/81=0.0**12345679**012345679012345679012345 | P 9 | 37=>10=>1[P=>D] |
| 1/82=0.0**12195**121951219512195121951219512 | P 5 | **18=>9** |
| **1/83**=0.0**12048192771084337349397590361445783132530**120 | | |
| | P 41 | **171=>9** |
| 1/84=0.01**190476**1904761904761904761904762 | P **6** | **27=>9** |
| 1/85=0.01**176470588235294**1176470588235294 | P 16 | **72=>9** |
| 1/86=0.01**1627906976744186046511**627906976 | P 21 | **99=>18=>9** |
| 1/87=0.01**149425287356321839080459770**1149 | P 28 | 123=>6 |
| 1/88=0.011**36**36363636363636363636363636 | P 2 | **9** |
| **1/89**=0.011**2359550561797752808988764044943820224719**101123 | | |
| | P 44 | **198=>18=>9 [P=>D]** |
| 1/90=0.011111111111111111111111111111111 | | all 1's |
| 1/91=0.01**10989**0109890109890109890109890 11 | P **6** | **27=>9** |
| 1/92=0.01**0869565217391304347826**086956521 | P 22 | **99=>18=>9** |
| 1/93=0.01**075268817204**3010752688172043011 | P 15 | **54=>9** |
| 1/94=0.01**06382978723404255319148936170212765957446808**5106382978 | | |
| | P 46 | **207=>9?** |
| 1/95=0.01**0526315789473684**210526315789473 | P 18 | **81=>9** |

1/96=0.0104166666666666666666666666666666                    all 6's

**1/97=0.01030927835051546391752577319587628865979381443298969072164948453
6082474226804123711340206185567**0103092783505154639

                                        P 96    **432=>9 [P=D-1]**


1/98=0.0**10204081632653061224489795918367346938775**510204081

                                        P 42    **189=>9**
1/99=0.010101010101010101010101010101010101           P 2     1
1/100=0.01                                             1
**1/101=0.0099**009900990099009900990099009900           P 4     **99=>18=>9**
**1/102=0.0098039215686274**509803921568627451           P 16    75=>12=>3
**1/103=0.009708737864077669902912621359223**0097       P 34    **153=>9**
1/104=0.009**615384**615384615384615384615384615846       P 6     **27=>9**
1/105=0.009**52380**9523809523809523809523809523809        P 6     **27=>9 [P=>D]**
1/106=0.009**433962264150**9433962264150943396            P 13    **54=>9**
**1/107=0.0093457943925233644859813084112149532710280373831775700934**

                                        P 53    **225=>9 [P=>D]**
1/108=0.009**259259259259259259259259259**2592592        P 3     16=>7
**1/109=0.00917431192660550458715596330275229357798165137614678899082568807
33944954128440366972477064220183486238532110**0917431**

                        P 108; **486=>9 [P=D-1]**
1/110=0.0090909090909090909090909090909090909           P 2     **9 [P=>D]**
1/111=0.0090090090090090090090090090090090009           P 3     **9 [P=>D]**
1/112=0.0089**285714**2857142857142857142857142857          P 6     **27=>9**
**1/113=0.008849557522123893805309734513274336283185840707964601769911504424778761061946902654867256637168141592920353982**300884

                        P 112; **504=>9 [P=D-1]**


## PI APPROXIMATION
355/**113** =
3.14159292035398230088495575221238938053097345132743362831858407079646017699115044247787610619469026548672566371681415929203539823**00884955752212389380530973451327433628318584070796460176991150442477876106194690265486725663716814159292035398230**08849557522123893805309734513274336283185
8407                                    P 112 **[P=D-1]**


1/114=0.008**771929824561403**5087719298245614           P 18    **81=>9**

1/115=0.**008695652173913043478260**8695652174    P 22    **99=>18=>9**

1/116=0.**008620689655172413793103448275**8621    P 28    **126=>9**

1/117=0.**008547**008547008    P **6**    24=>6

1/118=0.**00847457627**11**8644067796610169491525423728813559322033898305**08474
5762711

                                                 P 58    **261=>9**

1/119=0.**0084033613445378151260504201680672268907563025 21**008403361

                                                 P 48    **171=>9**

1/120=0.00833333    all 3's

1/121=0.**00826446280991735537190**0826446    P 22    **99=>18=>9? [P=>D]**

1/122=0.**0081967213114754098360655737704918032786885245901639344262295**081
96

                                                 P 60    **270=>9**

1/123=0.**008130**081300813    P 5    12=>3

1/124 = 0.00**80645161290322**580645161290322581    P 15    **54=>9**

1/125=0.008    8

1/126=0.00**793650**793650    P **6**    30=>3

**1/127=0.00787401574803149606299212598425196850393700**7874

                                                 P 42    **189=>18=>9**

1/128=0.0078125    23=>5

1/129=0.**0077519379844961240310**077519379844    P 21    **90=>9 [P=>D]**

1/130=0.00**76923**0769230769230769230769230769    P **6**    **27=>9**

**1/131=0.0076335877862595419847328244274809160305343511450381679389312977
09923664122137404580152671755725190839694656488549618320610687022900763**
3                                                P 130   **585=>18=>9 [P=D-1]**

1/132=0.007**5**75757575757575757575757575757575    P 2    12=>3

1/133=0.00**7518796992481203**007518796992481203    P 18    **81=>9**

1/134=0.00**7462686567164179104477611940298**507462686567

                                                 P 33    **153=>9**

1/135=0.00**740**740740740740740740740740740740    P 3    11=>2

1/136=0.00**7352941176470588235294117647058823**5294117647058823529411764705
882352941176470588235294117647058823529411176    P 32    **144=>9**

**1/137=0.00729927**007299270072    P 8    **36=>9**

1/138=0.00**724637681159420289855**0724637681159    P 22    102=>3

**1/139=0.00719424460431654676258992805753956834532374**100719

                                                 P 45    **204=>6**

1/140=0.00**714285**71428571428571428571428571    P **6**    **27=>9**

1/141=0.00**70921985815602836879432624113475177304964539**0070921

26

|  |  |  |
|---|---|---|
|  | P 46 | **207=>9** |
| 1/142=0.00**70422535211267605633**8028169014084507 | P 35 | **127=>9** |
| 1/143=0.00**6993**006993006993006993006993006 | P **6** | **27=>9** |
| 1/144=0.006944444444444444444444444444444444444 | all 4's |  |
| 1/145=0.00**689655172413793103448275862**06896 | P 28 | **126=>9 [P=>D]** |
| 1/146=0.00**68493150**68493150684931506849315068493151 | P 8 | **36=>9** |
| 1/147=0.00**68027210884353741496598639455782312925170**06802721088435374149659863945578231292517006802721088435374149659 | P 42 | **189=>18=>9** |
| 1/148=0.00**675675**67567567567567567567567567567 | P 3 | **18=>9** |
| **1/149=0.00**6711409395973154362416107382550335570469798657718120805369127516778523489932885906040268456375838926174496644295302013422818791946308724832214765**1006711409395** | P 148 | **666=>18=>9** |
|  |  | **[P=D-1]** |
| 1/150=0.006666666666666666666666666666666666 | All 6's |  |
| **1/151=0.00**66225165562913907284768211920529801324503311258278145695364238410596026490066225165562913907284**7** | P 75 | **306=>9** |
| 1/152 = 0.006**57894736842105263**15789473684211 | P 18 | **81=>9** |
| 1/153=0.00**653594771241830**0653594771241830006 | P 16 | 65=11=>2 |
| 1/154=0.00**6493506**49350649350649350649350649 | P **6** | **27=>9** |
| 1/155=0.00**645161290322580**6451612903225800645 | P 15 | **54=>9** |
| 1/156=0.00**641025**641025641025641025641025641 | P **6** | **18=>9** |
| **1/157=0.00**636942675159235668789808917197452229299363057324840764331210191082802547770700636942675159235668 | P 78 | **351=>9** |
| 1/158=0.00**63291139240**50632911392405063 | P 13 | **45=>9** |
| 1/159=0.00**628930817610**0628930817610062 | P 13 | 51=>6 |
| 1/160=0.00625 |  | 13=>4 |
| **1/161=0.00**621118012422360248447204968944099378881987577639751552795031055**9**00621 |  |  |
|  | P 66 | 317=>11=>2 |
|  |  |  |
| 1/162=0.00**6172839**50617283950617283950617 | P 9 | 41=>5 [**P=>D**] |
| **1/163=0.00**613496932515337423312883435582822085889570552147239263803680981595092024539877300613496932515337 | P 81 | **360=>9** |
| 1/164=0.00**60975**60975609756097560975609756097560975609 | P 5 | **27=>9** |
| 1/165=0.006060606060 | P 2 | 6 |
| 1/166=0.00**60240963855421686746987951807228915662650**602409 | P 41 |  |
| **1/167=0.00**5988023952095808383233532934131736526946107784431137724550898203592814371257485029940119760479041916167664670658682634730538922155688 |  |  |

**6227544910179640718562874251497**0059880239520958  P 166 **747=>18=>9**

**[P=D-1]**

1/168=0.005**952380**952380952380952380523809          P **6**       **27=>9**

1/169=0.**00591715976331360946745562130177514792899408284023668639053254437869822485207**10059171597633136094674556213017775  P 78    **351=>9**

1/170=0.00**5882352941176470**5882352941176470          P 16      **72=>9**

1/171=0.00**58479532163742690**058479532163742  P 18    **72=>9 [P=>D]**

1/172=0.00**58139534883720930232**5581395348837          P 21      **90=>9**

**1/173=0.0057803468208092485549132947976878612716763**00578034682080

P 43      **207=>9**

1/174=0.00**5747126436781609195402298**8505747  P 28    129=>3

1/175=0.005**714285714**          P **6**       **27=>9**

1/176=0.00**568**181818181          P 2       **9**

1/177=0.**00564971751412429378531073446327683615819209039548022598870**05649 7175141          P 58      **198=> 9**

1/178=0.00**56179775280898876404494382022471910112359550**561797752808988764 0449438          P 44      206=>8

**1/179=0.00558659217877094972067039106145251396648044692737430167597765363128491620111731843575418994413407821229050279329608938547486033519553072625698324022346368715083798882681564245810**0558659217877094972067039106145251396648044692737430167597765363128491620111731843575418994413407821229050279          P 178  **[P=D-1]**

1/180=0.005555555555          all 5's

## We Continue by Examining Prime Denominators

## First Few Prime Numbers

| 2 | **3** | **5** | **7** | **11** | 13 | 17 | 19 | **23** | **29** | 31 | 37 | **41** | **43** | **47** | 53 |
|---|---|---|---|---|---|---|---|---|---|---|---|---|---|---|---|
| **59** | **61** | 67 | 71 | 73 | **79** | **83** | 89 | **97** | **101** | 103 | 107 | 109 | **113** | 127 | **131** |
| **137** | 139 | **149** | **151** | 157 | 163 | **167** | **173** | 179 | 181 | **191** | 193 | 197 | 199 | | |
| 211 | **223** | 227 | 229 | 233 | **239** | **241** | 251 | **257** | **263** | 269 | 271 | **277** | **281** | 283 | |
| **293** | 307 | **311** | **313** | **317** | **331** | 337 | **347** | **349** | 353 | 359 | **367** | 373 | 379 | | |
| **383** | **389** | 397 | **401** | 409 | **419** | **421** | 431 | 433 | **439** | **443** | 449 | **457** | **461** | | |
| 463 | 467 | **479** | 487 | **491** | 499 | 503 | **509** | 521 | 523 | 541 | | | | | |

1/181=0.00**5524861878453038674033149171270718232044198895027624309392265193370165745856353591160220994475138121546961325966850828729281767955801104972375690607734806629834254143646408839779**00552486187845303867403314

9171270718232044198895027624309392265193370165745856353591160220994475138121546961                                    P 180 **[P=D-1]**

**1/191=0.00523560209424083769633507853403141361256544502617801047120418848167539267015706806282722513089**00523560209424083769633

P 95

**1/193=0.00518134715025906735751295336787564766839378238341968911917098445595854922279792746113989637305699481865284974093264248704663212435233160621761658031088082901554404145077720207253886010362694**300518134715025

P

**1/197=0.00507614213197969543147208121827411167512690355329949238578680203045685279187817258883248730964467**005076142131979695431

P 98     **[P=>D]**


**1/199=0.00502512562814070351758793969849246231155778894472361809045226130653266331658291457286432160804020100**50251256281407035

P 99

Not prime: 1/202=0.**00495**04950495                    P 4     18=>9   **[P=>D]**

Not Prime: 1/204=0.00**490196078431372**54901960784313725     P 16     69=>15=>6


Not Prime: 1/205 = 0.00487804878                    P 5     27=>**9**

**1/211=0.004739336492890995260663507109004739336492890995260663507109004739336492890995260663507109004739336492890995260663507109004739336492890995260663507109**004739336492890995260663507109004739336492890995260663

P 150

1/220 = 0.004545454                            P 2     **9**

**1/223=0.004484304932735426008968609865470852017937219730941704035874439461883408071748878923766816143497757847533632286995515695067264573991031390134529147982062780269058295964125560538116591928251121076233183856502242152466367713**004484304932735426008968609865470852017937219730941704035874439461

P 222 **[P=D-1]**

**1/227=0.004405286343612334801762114537444933920704845814977973568281938325991189427312775330396475770925110132158590308370**04405

P 112

**1/229=0.004366812227074235807860262008733624454148471615720524017467248908296943231441048034934497816593886462882096069868995633187772925764192139737991266375545851528384279475982532751091703056768558951965065502183406113537117903930131**00436681222707423580786026200873362445414847161572052401746720524017467

P 228 **[P=D-1]**

1/233=0.0042918454935622317596566523605150214592274678111587982832618025
75107296137339055793991416309012875536480686695278969957081545064377682
40343347639484978540772532188841201716738197424892703862660944206008583
69098712446351931330472103004291845493562231759656652360515021459227467
81115879828                                        P 232 **[P=D-1]**
1/239=**0.00418410**0418410041841004184100418         P 7       18=>**9**


1/241=**0.004149377593360995850622406639**0041493         P 30       135=>**9**
Not Prime: 1/246 = 0.0**040650**406504                  P 5    15=>6


1/251=**0.00398406374501992031872509960159362549800796812749**00398406374501
99203187250996015936254980079681274900398406374501
                                                   P 50
Not Prime: 1/252 = **0.00396825396825**                  P 6    33=>6


1/257=**0.0038910505836575875486381322957198443579766536964980544747081712
06225680933852140077821011673151750972762645914396887159533073929961089
49416342412451361867704280155642023346303501945525291828793774319066147
85992217898832684824902723735408560311284046692607**003891050583657587548
63813229571                                         P 256  **[P=D-1]**


1/263=**0.0038022813688212927756653992395437262357414448669201520912547528
51711026615969581749049429657794676806083650190114068441064638783269961
97718631178707224334600760456273764258555133079847908745247148288973384
03041825095057034220532319391634980988593155893536121673**003802281368821
29277566539                                         P 262  **[P=D-1]**


1/269=**0.0037174721189591078066914498141263940520446096654275092936802973
97769516728624535315985130111524163568773234200743494423791821561338289
96282527881040892193308550185873605947955390334572490706319702602230483
27137546468401486988847583643122676579925650557620817843866171**003717472
1189591078                                          P 268 **[P=D-1]**


1/271=**0.00369**0036900369003690036900369003690036900369         P 5       **18=>9**


1/277=**0.0036101083032490974729241877256317689530685920577617328519855595
66787**003610108303249097472                              P 69

30

1/280=0.003**5714285**714285714285714285714285          P **6**       27=>9

**1/281=0.0035587188612099644128113879**0035587          P 28      **126=>9**
Not Prime: 1/408=0.002**4509803921568627**450980392156863     P 16    75=>12=>3
**1/283=0.0035335689045936395759717314487632508833922261484098939929328621**
**9081272084805653710247349823321554770318021201413427561837455830388692**5
**7950530**0353356                                          P 140


1/287= 0.00**3484320557491289198606271777**003484          P 30


**1/293=0.0034129692832764505119453924914675767918088737201365187713310580**
**2047781569965870307167235494880546075085324232081911262798634812286689**4
**197952218430034129692832764505119453924914675**7     P 146
Not prime: 1/303=0.**0033**003300330033                    P 4    6
**1/307=0.0032573289902280130293159609120521172638436482084690553745928338**
**7622149837133550488599348534201954397394136807817589576547231270358306**1
**88925081433224755**7003257328990228013029315960912052
                              P 153
**1/311=0.0032154340836012861736334405144694533762057877813504823151125401**
**9292604501607717041800643086816720257234726688102893890675241157556270**0
**964630225080385852090**0321543408360128617363344405144
                              P 155
**1/313=0.0031948881789137380191693290734824281150159744408945686900958466**
**4536741214057507987220447284345047923322683706070287539936102236421725**2
**3961661341853035143769968051118210862619808306709265175718849840255591**0
**5431309904153354632587859424920127795527156549520766773162939297124600**6
**3897763578274760383386581469648562**3003194888178913738019169329073
                              P 312 **[P=D-1]**
**1/317=0.0031545741324921135646687697160883280757097791798107255520504731**
**8611987381703470**03154574132492113564668769**7**          P 79
1/330 = 0.00**30303030303**                               P 2     3
**1/331=0.0030211480362537764350453172205438066465256797583081570996978851**
**9637462235649546827794561933534743202416918429000302114803625377643504**53
**1722054380664652567975830815709969788519637462235649546827794561933534**7
**432024169184290**0030211480362537764350453172205438066465256797
                              P 220
**1/337=0.0029673590504451038575667655786350148367952522255192878338278931**
**75074183976261127596439169139465875370919881305637982195845697329376854**

31

5994065281899109792284866468842729970326409495548961424332344 2136498516
3204747774480712166172106824925816023738872403560830860534124 6290801186
9436201780415430267062314540059347181008902077151335311572700 29673

P 336 [P=D-1]

1/347=0.002881844380403458213256484149855907780979827089337175 7925072046
1095100864553314121037463976945244956772334293948126801152737 7521613832
8530259365994236311239193083573487031700288184438040345821325 6484149855
9077809798270893371757925072046109510086455331412103746397694 5244956772
3342939481268011527377521613832853025936599423631123919308357 3487031700
28818443                                   P 346 [P=D-1]
1/349=0.002865329512893982808022922636103151862464183381088825 2148997134
6704871060171919770773638968481375358166189111747851002865 3  P 116
1/353=0.002832861189801699716713881019830028328611898016997167 1388101983
0028328611898016997167138810198300283286118980169971671388101 9830028328
61189801699716                                   P 128 [P=>D]

1/359=0.002785515320334261838440111420612813370473537604456824 5125348189
4150417827298050139275766016713091922005571030640668523676880 2228412256
267409470752089136490250696378830083565459 6100278551532033

P 179 [P=>D]
1/367=0.002724795640326975476839237057220708446866485013623978 2016348773
8419618528610354223433242506811989100817438692098092643051771 1171662125
3405994550408719346049046321525885558583106267029972752043596 7302452316
0762942779291553133514986376021798365122615803814713896457765 6675749318
8010899182561307901907356948228882833787465940054495912806539 5095367847
41144414168937329 700272479564032697

P 366 <u>check</u>

1/373=0.002680965147453083109919571045576407506702412868632707 7747989276
1394101876675603217158176943699731903485254691689008042895442 3592493297
5871313672922252010723860589812332439678284182305630026809651 4745308310
9919571045576407506702412868632707774798927613941018766756032 1715817694
3699731903485254691689008042895442359249329758713136729222520 1072386058
9812332439678284182305630026809651474530831099195710455764075 0670241286
8632707774                                   P 372 [P=D-1]

1/379=0.0026385224274406332453825857519788918205804749340369393139841688
6543535620052770448548812664907651715039577836411609498680738786279 6833
7730870712401055408970976253298153034300791556728232189973614775725 5936
6754617414248021108179419525065963060686015831134564643799472295514 5118
7335092348284960422163588390501319261213720316622691292875989445910 2902
374670184696569920844327176781002638522427440633245382585751978891820 58
04749                                     P  378 <u>check</u>

1/383=0.0026109660574412532637075718015665796344647519582245430809399477
8067885117493472584856396866840731070496083550913838120104438642297 6501
3054830287206266318537859007832898172323759791122715404699738903394 2558
7467362924281984334203655352480417754569190600522193211488250652741 5143
6031331592689295039164490861618798955613577023498694516971279373368 1462
14099216710182767624020887728459530026109660574412532637075718015665796
3446                                     P  382 <u>check</u>

1/389=0.0025706940874035989717223650385604113110539845758354755784061696
6580976863753213367609254498714652956298200514138817480719794344473 0077
1208226221079691516709511568123393316195372750642673521850899742930 5912
5964010282776349614395886889460154241645244215938303341902313624678 6632
3907455012853470437017994858611825192802056555269922879177377892030 8483
29048843187660668380462724935732647814910025706940874035989717223650385
6041131                                     P  388 <u>check</u>

1/397=0.0025188916876574307304785894206549118387909319899244332493702770
7808564231738035264483627204030226700251889168765743073047858942065 4911
8387909319899244332493702770780856423173803526448362720403022670025 1889
1687657430730478589420654911838790931989924433249370277078085642317 3803
526448362720403022670025188916876574307304785894206549118387909319899924
                                     P
1/401=0.0024937655860349127182044887780548628428927680798004987531172069
8254364089775561097256857855361596009975062344139650872817955112219 4513
7157107231920199501246882793017456359102244389027431421446384039900 2493
76558603491271820448877805486284289276807980049875311720698254364089 77
                                     P
1/409=0.0024449877750611246943765281173594132029339853300733496332518337
4083129584352078239608801955990220048899755501222493887530562347188 2640
5867970660146699266503667481662591687041564792176039119804400977995 1100

33

2444987775061124694376528117359413202933985330073349633251833740831295

P

**1/419=0.00**2386634844868735083532219570405727923627684964200477326968973747016706443914081145584725536992840095465393794749403341288782816229116945107398568019093078758949880668257756563245823389021479713603818615751789976133651551312649164677804295942720763723150357995226730310262529832935560859188544152744630071599045346062052505966587112171837708830548926014319809069212410501193317422434367541766109785202863961813842482100238663484486873508353221957040572792362768496420047732696897374701670644391408114558472                                    P

**1/421=0.00**2375296912114014251781472684085510688836104513064133016627078384798099976247030878859857482185273159144893111638954869358669833729216152019**00**23752969121140142517814726840855106888361045130641330166270783847980997624703087885985748218527315914489311163895486935866983372921615201900023752969121140142517814726840855106888361045130641330166270783847980997624703087885985748218527315914489311163895486935866983372921615201900023752969121140142517814726840855106888361045130641330166270783847900**23752969121140142517814726840855106888361045130641330166270783** P

**1/431=0.00**23201856148491879350348027842227378190255220417633410672853828306264501160092807424593967517401392111368909512761020881670533642691415313225058004640371229698375870069605568445475638051044083526682134570765661252900**23201856148491879350348027842227378190255220417633410** P

**1/433=0.00**23094688221709006928406466512702078521939953810623556581986143187066974595842956120092378752886836027713625866050808314087759815242494226327944572748267898383371824480369515011547344110854503464203233256351039260969976905311778290993071593533487297921478060046189376443418013856812933025404157043879907621247113163972286374133949191685912240184757505773672055427251732101616628175519630484988452655889145496535796766743648960739030**0023094688221709006928406466512702078521939953810623556581986143187066974595842956120092378752886836027713625866050808314087759815242494226327944572748267898383371824480369515011547344110854503464203233256351039260969976905311778290993071593533487297921478060046189376443418013856812933025404157043879907621247113163972286374133949191685912240184757505773672055427251732101616628175519630484988452655889145496535796766743648960739030**0023094688221709006928406466512702078521939953810623556581986143187066974595841456535796766743648960739030**0023094688221709006928406466512702078521939953810623556581986143187066974595841456535796766743648960739030 P

**1/439=0.00**22779043280182232346241457858769931662870159453302961275626423690205011389521640091116173120728929384965831435079726651481

**06378132118451025056947608200455580865603644646924829157175398633**
**25740318906605922551252847380410**022779043280182232346241457858769
9316628701594533                                P


**1/443=0.0022573363431151241534988713318284424379232505643340857787**
**81038374717832957110609480812641083521444695259593679458239277652**
**37020316027088036117381489841986455981941309255079006772009029345**
**3724604966139954853273137697516930**022573363431151241534988713318284424379232505643340857787
844243792325056433                              P


**1/449= 0.0022271714922048997772828507795100222717**
                                   P 32   **144=>9**


1/451=0.**00221729490**022172949002217                P 10     **[P=>D]**

**1/457=0.0021881838074398249452954048140043763676148796498905908096280087**
**52735229759299781181619256017505470459518599562363238512035010940919037**
**199124726477024070**02188183807439824945295404814004376367614879649
                                   P


**1/461=0.0021691973969631236442516268980477223427331887201735357917570498**
**91540130151843817787418655097613882863340563991323210412147505422993492**
**40780911062906724511930585683297180043383947939262472885032537960954446**
**85466377440347071583514099783080260303687635574837310195227765726681127**
**98264642082429501084598698481561822125813449023861171366594360086767895**
**87852494577006507592190889370932754880694143167028199566160520607375271**
**14967462039045553145336225596529284164859**0021691973969631236442516268980
477223427331887201735357917570498          P


**1/463=0.0021598272138228941684665226781857451403887688984881209503239740**
**82073434125269978401727861771058315334773218142548596112311015118790496**
**760259179265658747300215982721382289416846652267818574514038876889848812**
**0950323974082073434125269978401727861771058315334773218142548596112311**
**0151187904967602591792656587473**0021598272138228941684665226781857451403
8876889                                      P


35

1/467=0.0021413276231263383297644539614561027837259100642398286937901498
9293361884368308351177730192719486081370449678800856531049250535 3319057
8158458244111349036402569593147751605995717344753747323340471092 0770877
9443254817987152034261241970021413276231263383297644539614561027 8372591
0064239828693790149892933618843683083511777301927194860813704496 7880085
6531049250535331905781584582441113490364025695931477516059957173 4475374
7323340471092077087794432548179871520342612419700214132762312633 8329764
4539614561027837259                                                    P

Not prime: 1/478= 0.00**20920**50209205020920              P 7      18=>9

1/479=0.0020876826722338204592901878914405010438413361169102296450939457
2025052192066805845511482254697286012526096033402922755741127348 6430062
6304801670146137787056367432150313152400835073068893528183716075 1565762
0041753653444676409185803757828810020876826722338204592901878914 4050104
3841336116910229645093945720250521920668058455114822546972860125 2609603
3402922755741127348643006263048016701461377870563674321503131524 0083507
3068893528183716075156576200417536534446764091858037578288100208 7682672
2338204592901878914405010438413361169102296450939457202505219206 6805845
5114822546972860125260960                                               P

1/491=0.0020366598778004073319755600814663951120162932790224032586558044
8065173116089613034623217922606924643584521384928716904276985743 3808553
9714867617107942973523421588594704684317718940936863543788187372 7087576
3747454175152749490835030549898167006109979633401221995926680244 3991853
3604887983706720977596741344195519348268839103869653767820773930 7535641
5478615071283095723014256619144602851323828920570264765784114052 9531568
2281059063136456211812627291242362525458248472505091649694501018 3299389
00203665987780040                                                      P

| 373 | 379 | **383** | **389** | 397 | **401** | 409 |
| **419** | **421** | 431 | 433 | **439** | **443** | 449 | **457** | **461** | 463 |
| 467 | **479** | 487 | **491** | 499 | 503 | **509** | 521 | 523 | 541 |

1/499=0.0020040080160320641282565130260521042084168336673346693386773547
0941883767535070140280561122244488977955911823647294589178356713 4268537
0741482965931863727454909819639278557114228456913827655310621242 4849699

36

3987975951903807615230460921843687374749498997995991983967935871743486
9739478957915831663326653306613226452905811623246492985971943887775551
1022044088176352705410821643286573146292585170340681362725450901803607
214428857715430861723446893787575150300601202404809619238476953907815631
2625250501002004                                   P 498  **[P=D-1]**

1/503=0.**00198807157057654075546719681908548707753479125248508946322067**59
44333996023856858846918489065606361829025844930417495029821073558648111
33200795228628230616302186878727634194831013916500994035785288270377733
59840954274353876739562624254473161033797216699801192842942345924453280
31809145129224652087475149105367793240556660039761431411530815109343936
38170974155069582504970178926441351888667992047713717693836978131212723
65805168986083499005964214711729622266401590457256461232604373757455268
3896620278330019880715**7**                         P 502  **[P=D-1]**

1/509=0.**001964636542239685658153241650294695481335952848722986247544**2043
22200392927308447937131630648330058939096267190569744597249508840864440
07858546168958742632612966601178781925343811394891944990176817288801571
70923379174852652259332023575638506876227897838899803536345776031434184
67583497053045186640471512770137524557956777996070726915520628683693516
69941060903732809430255402750491159135559921414538310412573673870333988
21218074656188605108055009823182711198428290766208251473477406679764243
6149312377210216110019646365422396856581532416502

                                              P 508  **[P=D-1]**
1/521=0.**00191938579654510556621880998080614203454894433781194433781190**01
91938579654510556621880998080614203454894433781190**0**

                                              P 61
1/523=0.**001912045889101338432122370936902485659655831739961759082217**9732
31357552581261950286806883365200764818355640535372848948374760994263862
33269598470363288718929254302103250478011472275334608030592734225621414
91395793499043977055449330783938814531548757170172084130019120458889101
38432122370936902485659655831739961759082217973231357552581261950286806
88336520076481835564053537284894837476099426386233269598470363288718929
25430210325047801147227533460803059273422562141491395793499043977055449
3307839388145315487571701720841**3**00191204588910133

                                              P 522  **[P=D-1]**

1/533= 0.**00187617260787992495309568480**3001876172  P 30

37

**1/541**=0.**001848428835489833641**4048059149722735674676524953789279112754158
964879852125693160813308687615526802218114602587800369685767097966672828
096118299445471349353049907578558225508317929759704251386321626617375
231053604436229205175600739371534195933456561922365988909426987060998151
571164510166358595194085027726432532347504621072088724584103512014787430
683918669131238447319778188539741219963031423290203327171903881700554
528650646950092421441774491682070240295748613678373382624768946395563770
79482439926062846580406654343807763401109057301293900**1848428835489833641
40                                                      P 540  [P=D-1]**

1/574=0.**00174216027874564459930313588850**0174216      P 30

1/861=0.**00116144018583042973286875725900**11614      P 30

1/902=0.**001108647450**11086474501108647450011086      P 10

Not prime: 1/717=0.**0013947**001394700      P 7    24=>6

1/936=0.001**068376**0683760683760683      P 6    30=>3

**1/1013**=0.000098716683119447186574531095755182625863770977295162882527147
087857847976307996051332675222112537018756169792694965449160908193484698
914116485686080947680157946692991115498519249753208292201382033563672
260612043435340572556762092793682132280355380059230**00098716683119**
                          P 253

1/1019=0.**000981354268891069676153091265947006869479882237487733071638861
629048086359175662414131501472031403336604514229636898920510304219823
3562315996074582924435721295387634936211972522080471050049067713444553483
807654563297350343473994111874386653581943081452404317958783120706575
073601570166830225711481844946025515210991167811579980372914622178606476
938174681059862610402355250245338567222767419038272816486751717369970559
371933267909715407262021589793915603532875368007850834151128557409224
730127576054955839057899901864573110893032384690873405299313052011776251
226692836113837095191364082433758586849852796859666339548577036310107948
969578017664376840039254170755642787046123650637880274779195289499509
3228655544651619234543670264965652600588812561334641805691854759568204121
687929342492639842983316977428851815505397448478900883218842001962708 53**

38

778213935230618253189401373895976447497546614327772325809617271835132482826300294406280667320902845927379784102060843964671246319921491658488714425907752698724239450441609421000981354268891069676153091265

P 1018  [P=D-1]

1/7841=0.000127534753220252518811376099987246524677974748118862390001275347532202525188113760999872465246779747481188 6   P 56   252=>9 [P=>D]


1/7901=0.00012656625743576762435134793064169092519934185546133400835337299076066320718896342235160106315656246044804455132261739020377167447158587520567016833312238957094038729274775344893051512466776357423110998607771168206556132135172762941399822807239589925325908112897101632704720921402354132388305277812935071509935451208707758511580812555372737628148335653714719655739779774712061764333628654600683457790153145171497278825465130996076446019491203645108214150107581318820402480698645741045437286419440577142133907100367042146563726110618908998860903683078091380837868624224781673205923300847993924819643083154031135299329198835590431590937855967599038096443488166054929755727123148968485001898493861536514365270218959625363877990127831920010125300594861409948107834451335274015947348436906720668269839260853056575117073788128085052524996835843564105809391216301733957726870016453613466649791165675230983419820275914441209973421085938488798886216934565244905708138210353119858245791671940260726490317681306163776737121883305910644222250348057207948360966966206809264650044298190102518668522971775724591823819769649411466902923680546766232122516137197823060372104796861156815592962916086571320086065055056321984558916592836349829135552461713707125680293633717250980888495127199088722946462473104670294899379825338564738640678395139855714466523224908239463335906847234527275028477407923047715479053284394380458169851917478800151879508922921149221617516770029110239210226553600810024047588912795848626756106821921275787874952537653461587140868244526009365903050246804201999746867485128464751297304138716618149601316289077331983293254018478673585622073155296797873686875079103910897354765219592456651056828249588659663333755220858119225414504493102138969750664472851537780027844576635868877357296544741172003543855208201493481837742057967345905581571952917352233894443741298569801290975825844829768383748892545247437033286925705606885204404505758764713327426907986330844196937096570054423490697380078471079610175927097835716997848373623591950386027085179091254271611188457157321857992659157068725477787621820022781926338438172383242627515504366535881533983040121503607138336919377294013416023288191368 18

39

12428806480192380711302366789014048854575370206302999620301227692697126
94595620807492722440197443361599797493988102771801037843310973294519681
05303126186558663460321478293886849765852423743829894950006328312871788
38121756739653208454625996709277306670041766864953803316035944817111758
00531578281230224022275661308695101885837235792937602835084166561194785
47019364637387672446525756233388178711555499303885584103278066067586381
47069991140361979496266295405644855081635236046070117706619415263890646
75357549677256043538792557904062776863688140741678268573598278698898873
56030882166814327300341728895076572585748639412732565498038223009745601
82255410707505379065941020124034932287052271864320972028857106695355018
35210732818630553094544994304518415390456904189343121123908366029616504
23996962409821541577015567649664599417795215795468927983799519048221744
08302746487786356157448424250094924693076825718263510947981268193899506
39159600050626502974307049740539172256676370079736742184533603341349196
30426528287558536894064042526262498417921782052904695608150866978863435
00822680673332489558283761549170991013795722060498671054296924439944310
84672826224528540691051765599291228958359701303632451588406530818883685
60941652955322111125174028603974180483483103404632325022149095051259334
26148588786229591190988482470573345146184027338311606125806859891153018
60523984305784077964814580432856600430325275281609922794582964181749145
67776230856853562840146816858625490444247563599544361473231236552335147
44968991266928236932033919756992785723326161245411973167953423617263637
51423870396152385773952664219719022908492595873940007593975446146057461
08087583850145551196051132768004050120237944563979243133780534109606378
93937476268826730793570434122263004682951525123402100999873433742564232
37564865206935830907480065814453866599164662700923933679281103657764839
89368434375395519554486773826097962283255284141247943298316668776104290
59612707252246551069484875332236425768890013922288317934438678648272370
58600177192760410074674091887102898367295279078597645867611694722187064
92849006454879129224148841918744462726237185166434628528034426022022528
79382356663713453993165422098468548285027211745348690039235539805087963
54891785849892418681179597519301354258954562713580559422857866092899632
95785343627388938109100113909631692190861916213137577521832679407669915
20060751803569168459688647006708011644095684090621440324009619035565118
33945070244272876851031514998101506138463485634729781040374636122009872
16807998987469940513859005189216554866472598405265156309327933173016073
91469434248829262118719149474750031641564358941906087836982660422731299
83546386533350208834324769016580179724085558790026578914061511201113783

06543475509429186178964688014175420832805973927350968231869383622326287
81166940893557777496519427920516390330337931907353499557018098974813314
77028224275408176180230350588533097076319453233767877483862802176939627
89520313884318440703708391342867991393494494367801544108340716365017086
44475382862928743197063662827490191115048728009112770535375268953297051
00620174661435261359321604860144285533476775091760536640931527654727249
71522592076952284520946715605619541830148082521199848120491077078850778
38248322997088976078977344639918997595241108720415137324389317807872421
21250474623465384128591317554739906340969497531957980002531325148715352
48702695861283381850398683710922668016706745981521326414377926844703202
12631312492089608910264523478040754334894317175041134033666624477914188
07745854955068978610302493355271484622199721554233641311226427034552588
27996456144791798506518162257942032654094418428047082647766105556258701
43019870902417415517023161625110745475256296671307429439311479559549424
12352866725730920136691558030629034299455765093026199215289203898240729
02164283002151626376408049613972914820908745728388811542842678142007340
84293127452221237817997721807366156182761675737248449563346411846601695
98784963928616630806227059865839767118086318187571193519807619288697633
21098595114542462979369700037969877230730287305404379192507277559802556
63840020250601189722819896215668902670548031894696873813441336539678521
70611315023414757625617010504999367168712821161878243260346791545374003
29072269332995823313504619668396405518288824199468421718769775977724338
69130489811416276420706239716491583343880521452980635362612327553474243
76661182128844450069611441589672193393241361852930008859638020503733704
59435514491836476395392988229338058473610935324642450322743956461207442
09593722313631185925832173142640172130110112643969117833185672699658271
10492342741425136058726743450196177699025439817744589292494620934058979
87596506771294772813567902797114289330464498164789267181369446905455005
69548158460954309581065687887609163397038349576003037590178458422984432
35033540058220478420453107201620048095177825591697253512213643842551575
74990507530692317428173648905201873180610049360840399949373497025692950
25946082774332362992026325781546639665865080369573471712441463105935957
47373750158207821794709530439184913302113656499177319326667510441716238
45082900898620427793950132894570307556005568915327173775471459308948234
40070877104164029869636754841159346918111631439058347044677888874825971
39602581951651689659536767497785090494874066573851411213770408809011517
52942665485381597266168839387419314010884698139476015694215922035185419
56714339956967472471839007720541703581825085432223769143146437159853183

14137450955575243640045563852676876344766485255031008733071763067966080
24300721427667383875458802683204657638273636248576129603847614226047335
78028097709150740412605999240602455385394253891912416149854448803948867
23199594987976205543602075686621946589039362106062523731173269206429565
87773699531704847487659789900012656625743576762435134793064169092519934
18554613340083533729907606632071889634223516

P 7900  [P=D-1]

1/7919=0.0001262785705265816390958454350296754640737466851875236772319 73
73405733047101906806414951382750347266068948099507513574946331607526202
80338426569011238792776865765879530243717641116302563454981689607273645
66233110241192069705770930673064780906680136380856168708170223513069832
04950119964642000252557141053163278191690870059350928147493370375047354
46394746811466094203813612829902765500694532137896199015027149892663215
05240560676853138022477585553731531759060487435282232605126909963379214
54729132466220482384139411541861346129561813360272761712337416340447026
13966409900239929284000505114282106326556383381740118701856294986740750
09470892789493622932188407627225659805531001389064275792398030054299785
32643010481121353706276044955171107463063518120974870564465210253819926
75842909458264932440964768278823083722692259123626720545523424674832680
89405227932819800479858568001010228564212653112766763480237403712589973
48150018941785578987245864376815254451319611062002778128551584796060108
59957065286020962242707412552089910342214926127036241949741128930420507
63985351685818916529864881929536557646167445384518247253441091046849349
66536178810455865639600959717136002020457128425306225533526960474807425
17994696300037883571157974491728753630508902639222124005556257103169592
12021719914130572041924485414825104179820684429852254072483899482257860
84101527970703371637833059729763859073115292334890769036494506882182093
69869933072357620911731279201919434272004040914256850612451067053920949
61485035989392600075767142315948983457507261017805278444248011112514206
33918424043439828261144083848970829650208359641368859704508144967798964
51572168203055941406743275666119459527781462305846697815380729890137 64
36418739739866144715241823462558403838868544008081828513701224902134107
84189922970071978785200151534284631897966915014522035610556888496022225
02841267836848086879656522288167697941659300416719282737719409016289935
59792903144336406111882813486551332238919055436292461169339563076145978
02752872837479479732289430483646925116807677737088016163657027402449804
26821568379845940143957570400303068569263795933830029044071221113776992

42

04445005682535673696173759313044576335395883318600833438565475438818032
57987119585806288672812223765626973102664477838110872584922338679126152
29195605550574567495895946457886096729385023361535547417603232731405480 4
89960853643136759691880287915140800606137138527591867660058088142442227
55398408890011365071347392347518626089152670791766637201666877130950877
63606515974239171612577345624447531253946205328955676221745169844677358
25230458391211011491349917918929157721934587700467230710948352064654628
10960979921707286273519383760575830281601212274277055183735320116176284
88445510796817780022730142694784695037252178305341583533274403333754261
90175527213031948478343225154691248895062507892410657911352443490339689
35471650460916782422022982699835837858315443869175400934461421896704129
30925621921959843414572547038767521151660563202424548554110367470640232
35256976891021593635560045460285389569390074504356610683167066548806667
50852380351054426063896956686450309382497790125015784821315822704886980
67937870943300921833564844045965399671675716630887738350801868922843793
40825861851243843919686829145094077535042303321126404849097108220734941
28046470513953782043187271120090920570779138780149008713221366334133097
61333501704760702108852127793913372900618764995580250031569642631645409
77396135875741886601843667129688091930799343351433261775476701603737845
68758681651723702487687839373658290188155070084606642252809698194216441
46988256092941027907564086374542240181841141558277560298017426442732668
26619522667003409521404217704255587826745801237529991160500063139285263
29081954792271751483773203687334259376183861598686702866523550953403207
47569137517363303447404975375678747316580376310140169213284505619396388
43288293976512185882055815128172749084480363682283116555120596034852885
46533653239045334006819042808435408511175653491602475059982321000126278
57052658163909584543502967546407374668518752367723197373405733047101906
80641495138275034726606894809950751357494633160752620280338426569011238
79277686576587953024371764111630256345498168960727364566233110241192069
70577093067306478090668013638085616870817022351306983204950119964642000
25255714105316327819169087005935092814749337037504735446394746811466094
20381361282990276550069453213789619901502714989266321505240560676853138
02247758555373153175906048743528223260512690996337921454729132466220482
38413941154186134612956181336027276171233741634044702613966409900239929
28400050511428210632655638338174011870185629498674075009470892789493622
93218840762722565980553100138906427579239803005429978532643010481121353
70627604495517110746306351812097487056446521025381992675842909458264932
44096476827882308372269225912362672054552342467483268089405227932819800

4798585680010102285642126531127667634802374037125899734815001894178557898724586437681525445131961106200277812855115847960601085995706528602096224270741255208991034221492612703624194974112893042050763985351685818916529864881929536557646167445384518247253441091046849349665361788104558656396009597171360020204571284253062255335326960474807425179946963000378835711579744917287536305089026392221240055562571031695921202171991413057204192448541482510417982068442985225407248389948225786084101527970703371637833059729763859073115292334890769036494506882182093698699330723576209117312792019194342720040409142568506124510670539209496148503598939260007576714231594898345750726101780527844424801111251420633918424043439828261144083848970829650208359641368859704508144967798964515721682030559414067432756661194595277181462305846697815380729890137643641873973986614471524182346255840383886854400808182851370122490213410784189922970071978785200151534284631897966915014522035610556888496022250284126783684808687965652228816769794165930041671928273771940901628993559792903144336406111882813486551332238919055436292461169339563076145978027528728374794797322894304836469251168076777370880161636570274024498042682156837984594014395757040030306856926379593383002904407122111377699204445005682535673696173759313044576335539588331860083343856547543881803257987119585806288672812223765626973102664477838110872584922338679126152291956055057456749589594645788609672938502336153554741760323273140548048996085364313675969188028791514080060613713852759186766005808811424422275539840889001136507134739234751862608915267079176663720166687713095087763606515974239171612577345624447531253946205328955676221745169844677358252304583912110114913499179189291577219345877004672307109483520646546281096097992170728627351938376057583028160121227427705518373532011617628488445510796817780022730142694784695037252178305341583533274403333754261901755272130319484783432251546912488950625078924106579113524434903396893547165046091678242202298269983583785831544386917540093446142189670412930925621921959843414572547038767521151660563202424548554110367470640232352569768910215936355600454602853895693900745043566106831670665488066675085238035105442606389695668645030938249779012501578482131582270488698067937870943300921833564844040459653996716757166308877383508018689228437934082586185124384391968682914509407753504230332112640484909710822073494128046470513953782043187271120090920570779138780149008713221366334133097613335017047607021088521277939133729006187649955802500315696426316454097739613587574188660184366712968809193079934335143326177547670160373784568758681651723702487687839373658290188155070084066422528096981942164414698825609294102700

907564086374542240181841141558277560298017426442732668266195226670034090
521404217704255587826745801237529991160500063139285263290819547922717514
837732036873342593761838615986867028665235509534032074756913751736330344
740497537567874731658037631014016921328450561939638843288293976512185
882055815128172749084480363682283116555120596034852885465336532390453340
068190428084354085111756534916024750599823210001262785705265816390958454
35029675464073746685187523677231973734055
                                                    P 7918  **[P=D-1]**

## Multiples of Inverse Doubly Prime Number 7

1/7 yields a six digits repeating sequence: namely, **1 4 2 8 5 7.** We write this mathematically as

RB(1/7) = **1 4 2 8 5 7.** The sum of this Repeat Block is 27 with a SDQ of 9. We therefore write this as SDQ(RB(1/7)) = 1 4 2 8 5 7 = 27=>9.

[Three female and three male; 2 4 8 and 1 5 7]

Let us examine the Repeat Block of the inverse of 7, when it is multiplied by consecutive digits. Each **Repeat Block** has the same 6 digits **142857.**

| | | |
|---|---|---|
| 1/7 = 0.**142857**142857... | P 6 | 27=>9 |
| 2/7 = 0.**285714**285714285714285714285714285714**29** | P 6 | 27=>9 |
| 3/7 = 0.**428571**428571428571428571428571142857143 | P 6 | 27=>9 |
| 4/7 = 0.**571428**571428571428571428571428571428571428**57** | P 6 | 27=>9 |
| 5/7 = 0.**714285**714285714285714285714285714285714285**71** | P 6 | 27=>9 |
| 6/7 = 0.**857142**857142857142857142857142857142857142**86** | P 6 | 27=>9 |
| 7/7 = 1 | | |
| 8/7 = 1.**142857**142857142857142857142857142857142857**71** | P 6 | 27=>9 |
| 9/7 = 1.**285714**285714285714285714285714285714**143** | P 6 | 27=>9 |
| 10/7 = 1.**428571**428571428571428571428571428571428571**4** | P 6 | 27=>9 |
| 11/7 = 1.**571428**571428571428571428571428571428571428**86** | P 6 | 27=>9 |
| 12/7 = 1.**714285**714285714285714285714285714285714285**7** | P 6 | 27=>9 |
| 13/7 = 1.**857142**857142857142857142857142857142857142**1429** | P 6 | 27=>9 |
| 14/7 = 14/7 = 2 | | |

## Doubling the Denominator of 1/7
### Denominator SDQ

| | P | SDQ |
|---|---|---|
| 1/7 = 0.**142857**142857142857142857142857142857142**5714** | 7=>7 | 27=>**9** |
| 1/14 = 0.0**714285**714285714285714285714285714**28571** | 14=>5 | 27=>**9** |
| 1/28 = 0.03**571428**571428571428571428571428571**4286** | 28=>10=>1 | 27=>**9** |
| 1/56 = 0.017**857142**857142857142857142857142857**143** | 56=>11=>2 | 27=>**9** |

1/112 = 0.0089**285714**285714285714285714285714          112=>4          27=>**9**

1/224 = 0.0044**64285714**285714285714285714285714          224=>8          27=>**9**

1/448 = 0.0022321**42857**142857142857142857142**9**          448=>16=>7          27=>**9**

1/896 = 0.0011160**7142857**142857142857142857142**9**          896=>23=>5          27=>**9**

1/1792 = 0.000558035**714285**714285714285714285**71**          1792=>19=>1          27=>**9**

1/3584 = 0.0002790178**5714285**714285714285714286          3584=>20=>2          27=>**9**


Notice the pattern of the Denominator SDQ, starting with 1/7, namely: 7, 5, 1, 2, 4, 8, …

Which are exactly the same digits as in the Repeat Block of 1/7.


Every doubling of the denominator of 1/7 produces the **same Repeat Block** pattern.


**Halving the Denominator of 1/7**

|  | Denominator SDQ | RB SDQ |
|---|---|---|
| 1/7 = 0.**142857**142857142857142857142857 | 7=>7 | 27=>**9** |
| 1/3.5 = 0.**2857142857**142857142857142857142**9** | 35=>8 | 27=>**9** |
| 1/1.75 = 0.**5714285714**2857142857142857142857 | 175=>13=>4 | 27=>**9** |
| 1/0.875 = 1.**142857**142857142857142857142**571** | 875=>20=>2 | 27=>**9** |

1/ 0.4375 = 2.**2857142857**142857142857142857143          4375=>19=>10=>127=>**9**

1/0.21875 = 4.**5714285714**285714285714285714286          21875=>23=>5          27=>**9**

1/0.109375 = 9.**142857**142857142857142857142**571**          109375=>25=>7          27=>**9**

1/0.0546875 = 18.**2857142857**142857142857142857**14**          546875=>35=>8          27=>**9**

1/0.02734375 = 36.**5714285714**2857142857142857142**9**          2734375=>31=>4          27=>**9**

1/0.013671875 = 73.**142857**142857142857142857142857          13671875=>38=>11=>2;

                                                                                         27=>**9**

Every halving of the denominator of 1/7 produces the **same Repeat Block** pattern.

Consider the sequence of 1 doubled, thus 1  2  4  8  16  32 with the SDQ values: **1 2 4 8 7 5**.

Of the pattern: $2^n$

$2^{-9}$  $2^{-8}$  $2^{-7}$  $2^{-6}$  $2^{-5}$  $2^{-4}$  $2^{-3}$  $2^{-2}$  $2^{-1}$  $2^0$  $2^1$  $2^2$  $2^3$  $2^4$  $2^5$  $2^6$  $2^7$  $2^8$  $2^9$

This is the pattern that was applied to 1/7. Each term is given by

$(1/7) (2^{-9}$  $2^{-8}$  $2^{-7}$  $2^{-6}$  $2^{-5}$  $2^{-4}$  $2^{-3}$  $2^{-2}$  $2^{-1}$  $2^0$  $2^1$  $2^2$  $2^3$  $2^4$  $2^5$  $2^6$  $2^7$  $2^8$  $2^9 )$


Try (1/11), (1/13), (1/17), (1/19), 21, (1/23), (1/29), (1/31), and (1/71) times this pattern.

## Multiples and Inverse Multiples of 7 in Ascending Order

| Number | SDQ (denominator) | Repeat Block | First Digit of RB | SDQ (RB) |
|---|---|---|---|---|
| 1/3584 | 2 | 857142 | 8 | 27=>9 |
| 1/1792 | 1 | 571428 | 5 | 27=>9 |
| 1/896 | 5 | 714285 | 7 | 27=>9 |
| 1/448 | 7 | 142857 | 1 | 27=>9 |
| 1/224 | 8 | 428571 | 4 | 27=>9 |
| 1/112 | 4 | 285714 | 2 | 27=>9 |
| 1/56 | 2 | 857142 | 8 | 27=>9 |
| 1 /28 | 1 | 571428 | 5 | 27=>9 |
| 1/14 | 5 | 714285 | 7 | 27=>9 |
| | | | | |
| 1/7 | 7 | 142857 | 1 | 27=>9 |
| | | | | |
| 1/3.5 | 8 | 285714 | 2 | 27=>9 |
| 1/1.75 | 4 | 571428 | 5 | 27=>9 |
| 1/0.875 | 2 | 142857 | 1 | 27=>9 |
| 1/ 0.4375 | 1 | 285714 | 2 | 27=>9 |
| 1/0.21875 | 5 | 571428 | 5 | 27=>9 |
| 1/0.109375 | 7 | 142857 | 1 | 27=>9 |
| 1/0.0546875 | 8 | 285714 | 2 | 27=>9 |
| 1/0.02734375 | 4 | 571428 | 5 | 27=>9 |
| | | | | |
| 1/0.013671875 | 2 | 142857 | 1 | 27=>9 |

**Every multiple and inverse multiple of 7 contains exactly one set of the same RB 6 digits 142857 in different order.**

We note that the product of the sequence **142857** = 2240 => 8.

## Multiples of Inverse Doubly Prime Number 11 – P 2

| | | |
|---|---|---|
| 1/11 = 0.09090909090909090909090909090909 | | P 2 |
| 2/11 = 0.18181818181818181818181818181818 | | P 2 |
| 3/11 = 0.27272727272727272727272727272727 | | P 2 |
| 4/11 = 0.36363636363636363636363636363636 | | P 2 |
| 5/11 = 0.45454545454545454545454545454545 | | P 2 |

6/11 = 0.5454545454545454545454545454545                           P 2
7/11 = 0.6363636363636363636363636363636                           P 2
8/11 = 0.7272727272727272727272727272727                           P 2
9/11 = 0.8181818181818181818181818181818                           P 2
10/11 = 0.9090909090909090909090909090909                          P 2
11/11 = 1
12/11 = 1.0909090909090909090909090909090                          P 2
13/11 = 1.1818181818181818181818181818181                          P 2

## Inverse Multiples of 11 in Ascending Order - P 2                 SDQ RB

1/2816 = 0.0003551**1363636363636363636363636**363636                36=>9
1/1408 = 0.0007102**2727272727272727272727272727**27                27=>9
1/704 = 0.0014204**5454545454545454545454545454**545                45=>9
1/352 = 0.0028409**0909090909090909090909090909**091                09=>9
1/176 = 0.0056818**1818181818181818181818181818**181                81=>9
1/88 = 0.0113**6363636363636363636363636363636**36                  36=>9
1/44 = 0.022**2727272727272727272727272727272**72                   27=>9
1/22 = 0.04**5454545454545454545454545454545**45                    45=>9
**1/11 = 0.09**090909090909090909090909090909090                    09=>9
1/5.5 = 0.**18**1818181818181818181818181818181818                   18=>9
1/2.75 = 0.**36**363636363636363636363636363636363                   36=>9
1/1.375 = 0.**72**727272727272727272727272727272727                  72=>9
1/0.6875 = 1.**45**454545454545454545454545454545455                 45=>9

1/0.34375 = 2.**90**909090909090909090909090909090909                90=>9
1/0.171875 = 5.**81**8181818181818181818181818181818                 81=>9
1/0.0859375 = 11.**63**6363636363636363636363636363364                63=>9
1/0.04296875 = 23.**27**2727272727272727272727272727272              27=>9

Every doubling and halving of the denominator of 1/11 produces the **same Repeat Block** pattern.

Although each set of 2 digits has an SDQ of 9, the digits vary from fraction to fraction.

## Multiples of Inverse of Prime Number 13 - P 6

                                                                    SDQ RB
1/13 = 0.**076923**076923076923076923076923307692307                27=>9

2/13 = 0.**153846**153846153846153846153846615                27=>9
3/13 = 0.**230769**230769230769230769230769923                27=>9
4/13 = 0.**307692**307692307692307692307692923                27=>9
5/13 = 0.**384615**384615384615384615384615384615538                27=>9

6/13 = 0.**461538**461538461538461538461538461538846                27=>9
7/13 = 0.**538461**538461538461538461538461538461538154                27=>9
8/13 = 0.**615384**615384615384615384615384615384615538462                27=>9
9/13 = 0.**692307**692307692307692307692307692307692307769                27=>9
10/13 = 0.**769230**769230769230769230769230769230769230307                27=>9
11/13 = 0.**846153**846153846153846153846153846153846153538                27=>9
12/13 = 0.**923076**923076923076923076923076923076923076692                27=>9
13/13 = 1
14/13 = 1.**076923**076923076923076923076923076923231                27=>9
15/13 = 1.**153846**153846153846153846153846153846462                27=>9
16/13 = 1.**230769**230769230769230769230769230769692                27=>9

## Multiples and Inverse Multiples of 13 in Ascending Order

1/208 = 0.0048**076923**076923076923076923076923076923               27=>9
1/104 = 0.0096**153846**153846153846153846153846153846               27=>9
1/52 = 0.019**230769**230769230769230769230769230769               27=>9
1/26 = 0.038**461538**461538461538461538461538461538               27=>9
**1/13 = 0.076923**076923076923076923076923076923076               **27=>9**
1/6.5 = 0.**153846**153846153846153846153846153846615               27=>9
1/3.25 = 0.**307692**307692307692307692307692307692231               27=>9

Two sets of the same 6 Repeat Block digits; **076923** and **153846**.
**[Both sets have three female and three male: 076923** has 0 2 6 and 3 7 9, while
**153846** contains 4 6 8 and 1 3 5.]

Every doubling and halving of the denominator of 1/13 produces the **same**
**Repeat Block** 6 pattern.

**Inverse of the Prime Number 17 - P 16**
**Multiples and Inverse Multiples of 17 in Ascending Order RB SDQ**
1/136 = 0.0073**52941176470588**2352941176470588                72=>9
1/68 = 0.014**70588235294117**6470588235294118                72=>9

49

1/34 = 0.0**2941176470588235**2941176470588235          72=>9

**1/17** = **0.0588235294117647**05882352941176471          72=>9

1/8.5 = 0.**1176470588235294**11764705882352294          72=>9

1/4.25 = 0.**2352941176470588**2352941176470588          72=>9

1/2.125 = 0.**4705882352941176**4705882352941176          72=>9

Although each set of 16 digits has an SDQ of 9, the digits vary from fraction to fraction.

1/17 = 0.**0588235294117647**05882352941176471          72=>9

2/17 = 0.**1176470588235294**11764705882352941          72=>9

3/17 = 0.**1764705882352941**17647058823529411          72=>9

4/17 = 0.**2352941176470588**2352941176470588          72=>9

5/17 = 0.**2941176470588235**2941176470588235          72=>9

6/17 = 0.**3529411764705882**3529411764705882          72=>9

7/17 = 0.**4117647058823529**4117647058823529          72=>9

8/17 = 0.**4705882352941176**4705882352941176          72=>9

9/17 = 0.**5294117647058823**5294117647058823          72=>9

10/17 = 0.**5882352941176470**5882352941176471          72=>9

11/17 = 0.**6470588235294117**6470588235294118          72=>9


One set of the same 16 repeating digits, **0588235294117647**.

The frequencies of the digits are: one 0, two 1, two 2, one 3, two 4, two 5, one 6, two 7, two 8 and one 9.

[Eight female and eight male]


**Multiples and Inverse Multiples of 19 in Ascending Order - P 18**


                                                                                **SDQ RB**

1/152 = 0.006**5789473684210526**315789473684211          81=>9

1/76 = 0.01**3157894736842105**263157894736842          81=>9

1/38 = 0.02**6315789473684210**526315789473684          81=>9

**1/19 = 0.0526315789473684210**52631578947368          81=>9

1/9.5 = 0.**1052631578947368421**0526315789474          81=>9

1/4.75 = 0.**2105263157894736842**1052631578947          81=>9

1/2.375 = 0.**4210526315789473684**2105263157895          81=>9

One set of the same repeating 18 digits, **052631578947368421**.

1/19 = 0.**0526315789473684210**52631578947368

2/19 = 0.**1052631578947368421**0526315789474

3/19 = 0.**1578947368421052631**5789473684211

4/19 = 0.**210526315789473684**21052631578947

**[One 0, two 1, two 2, two 3, two 4, two 5, two 6, two 7, two 8 and one 9]**
[Nine female and nine male]

One set of the same repeating 18 digits, **052631578947368421**.

**Multiples and Inverse Multiples of 21 in Ascending Order – P 6**

|  | SDQ RB |
|---|---|
| 1/336 = 0.0029**761904**76190476190476190476190476190**5** | 27=>9 |
|  |  |
| 1/168 = 0.0059**52380**95238095238095238095238095238**1** | 27=>9 |
| 1/84 = 0.011**904761**9047619047619047619047619047**62** | 27=>9 |
| 1/42 = 0.02**380952380**9523809523809523809523809**524** | 27=>9 |
| **1/21 = 0.0476190476**190476190476190476190476190**48** | 27=>9 |
| 1/10.5 = 0.0**952380952380**95238095238095238095238095 | 27=>9 |
| 1/5.25 = 0.**1904761904761**9047619047619047619047619 | 27=>9 |
| 1/2.625 = 0.**38095238095238**095238095238095238095238 | 27=>9 |
| 1/1.3125 = 0.**76190476190476**190476190476190476190476 | 27=>9 |

Two sets of the same 6 repeating digits; **047619** and **095238**.
1/21 = 0.**047619**047619047619047619047619047619**048**
2/21 = 0.**095238**095238095238095238095238095238095
3/21 = 0.**142857**142857142857142857142857142857714
4/21 = 0.**190476**190476190476190476190476190476619

Two sets of the same 6 repeating digits; **047619** and **095238**.
[Each set consists of three female and three male]

| | P 22 | 99=18=>9 |
|---|---|---|
| **1/23= 0.0434782608695652173913**04347826087 | P 22 | 99=18=>9 |
| **2/23= 0.0869565217391304347826**08695652174 | P 22 | |
| **3/23= 0.1304347826086956521739**1304347826 | P 22 | |
| **4/23= 0.1739130434782608695652**17739130435 | P 22 | |
| **5/23= 0.2173913043478260869565**2173913043 | P 22 | |
| **6/23= 0.2608695652173913043478**2608695652 | | |
| **7/23= 0.3043478260869565217391**3043478261 | | |
| **8/23= 0.3478260869565217391304**3478260869 | | |
| **9/23= 0.3913043478260869565217**3913043478 | | |
| **10/23=0.4347826086956521739130**43478260869 | | |

**11/23=0.47826086956521739130434**782608696
**12/23=0.52173913043478260869565**217391304
**13/23=0.56521739130434782608695**652173913


**[Two 0, two 1, two 2, three 3, two 4, two 5, three 6, two 7, two 8, two 9]**


**Multiples and Inverse Multiples of doubly prime 23 in Ascending Order**


1/92 = 0.010**8695652173913043478261**086956522
1/46 = 0.02**17391304347826086956521**73913043
**1/23 = 0.0434782608695652173913**04347826086        99=>18=>9
1/11.5 = 0.**0869565217391304347826**08695652174
1/5.75 = 0.**17391304347826086956521**739130434
One set of the same 22 repeating digits, **0434782608695652173913**.
Two 0, two 1, two 2, three 3, two 4, two 5, three 6, two 7, two 8, two 9.
[11 female and 11 male]


**Each Repeat Block contains two of each digit and three each of the digits 3 and 6.**


| | | |
|---|---|---|
| 1/26 = 0.0**384615**38461538461538461538461538 | P 6 | 27=>9 |
| 2/26 = 0.**076923**0769230769230769230769230**77** | P 6 | 27=>9 |
| 3/26 = 0.1**153846**15384615384615384615384**62** | P 6 | 27=>9 |
| 4/26 = 0.1**53846**15384615384615384615384615 | P 6 | 27=>9 |
| 5/26 = 0.1**923076**9230769230769230769230769 | P 6 | 27=>9 |
| 6/26 = 0.**230769**2307692307692307692307692**3** | P 6 | 27=>9 |
| 7/26 = 0.2**692307**6923076923076923076923**07** | P 6 | 27=>9 |
| 8/26 = 0.**307692**3076923076923076923076923**1** | P 6 | 27=>9 |


| | | |
|---|---|---|
| **1/29=0.034482758620689655172413793**103448 | P 28 | 126=>9 |
| **2/29=0.068965517241379310344827586**206897 | P 28 | 126=>9 |
| **3/29=0.10344827586206896551724137931**1034 | P 28 | 126=>9 |


| | | |
|---|---|---|
| **27/29=0.93103448275862068965517241379**31 | P 28 | 126=>9 |

**[Two 0, and 9; three 1, 2, 3, 4, 5, 6, 7, 8]**

**1/31=0.032258064516129**032258064516129032          P 15    54=>9

**2/31=0.064516129032258**064516129032258064          P 15    54=>9

**3/31=0.096774193548387**096774193548387096          P 15    81=>9

**4/31=0.129032258064516**129032258064516129          P 15   54 =>9

**5/31=0.161290322580645**16129032258064516

**6/31=0.193548387096774**19354838709677419

**7/31=0.225806451612903**22580645161290323

**8/31=0.258064516129032**25806451612903226

**9/31=0.290322580645161**29032258064516129

**10/31=0.322580645161290**32258064516129032

**11/31=0.354838709677419**35483870967741935

**12/31=0.387096774193548**38709677419354839

**13/31=0.419354838709677**41935483870967741

**14/31=0.451612903225806**45161290322580645

**15/31=0.483870967741935**48387096774193548

**16/31=0.516129032258064**51612903225806451

**17/31=0.548387096774193**54838709677419355


**30/31=0.967741935483870**96774193548387096


## Two Repeat Block patterns:
[Two 0, two 1, three 2, one 3, one 4, two 5, two 6, NO 7, one 8, one 9]
[One 0, one 1, NO 2, two 3, two 4, one 5, one 6, three 7, two 8, two 9]


**1/37=0.027**027027027027027027027027027027          P 3      9


**1/41=0.02439**024390243902439024390243902439024          P 5     18=>9


**1/43=0.023255813953488372093**023255813953          P 21    90=>9


**1/47=0.0212765957446808510638297872340425531914893617**021276

                                                P 92

47/51=0.**921568627450980**39215686274509803          P 16    75=>12=>3

50/51=0.**980392156862745**09803921568627451          P 16    75=>12=>3

**1/71=0.01408450704225352112676056338028169**0140    P 35    **126=>9**


2/71=0.**028169014084507042253521126760563380**281          P 35

4/71=0.**05633802816901408450704225352112676**0563     P 35

8/71=0.**1126760563380281690140845070422535**21126 P 35

**1/113 =**

0.**0088495575221238938053097345132743362831858407079646017699115044247787610619469026548672566371681**41592920353982300884955     P 112

**16/113=**

0.**141592920353982300884955752212389380530973451327433628318584070796460176991150442477876106194690265486725663716**814159292     P 112

**π**

**π Approximation**

**355/113 = 3 + 16/113 =**

3.**14159**2920353982300884955752212389380530973451327433628318584070796460176991150442477876106194690265486725663716814159292035398230088495575221238938053097345132743362831858407079646017699115044247787610619469026548672566371 68

**P 112**

**π Actual**

3.**14159**265358979323846264338327950288419716939937510582097494459230781640628620899862803482534211706798214808651328230664709384460955058223172535940812848111745028410270193852110555964462294895493038196442881097566593344612847564823378678316527120190914564856692346034861045432664821339360726024914127372458700660631558817488152092096282925409171536436789259036001133053054882046652138414695194151160943305727036575959195309218611738193261179310511854807446237996274956735188575272489122793818301194912983367336244065664308602139494639522473719070217986094370277053921717629317675238467481846766940513200056812714526356082778577134275778960917363717872146844090122495343014654958537105079227968925892354201995611212902196086403441815981362977477130996051870721134999999837297804995105973173281609631859502445945534690830264252230825334468503526193118817101000313783875288658753320838142061717766914730359825349042875546873 1

1595628638823553787593751957781857780532171226806613001927876611195909216420198+

**No Repeat Block [?]**

54

1/1!=1

½!=.5

**1/3**!=1/6=0.16666666666666666        all 6's

¼!=1/24=0.041666666666666666    all 6's

**1/5!=1/120=0.00833333333333333**    all 3's

1/6!=1/720=0.0013888888888888    all 8's

**1/7!**=1/5,040=0.0001**9841269**841269    P 6    30=>3

1/8!=1/40,320=0.000024**801587**01587    P 6    24=>6

1/9!=1/362,880=0.0000027**557319223985890652**55731    P 18    89=>17=>8

1/10!=1/3,628,800=0.00000027**55731922398589065**255731    P 18    89=>17=>8

**1/11**!=1/39,916,800=0.000000002**5052108385441718**775052    P 18    76=>13=>4

1/12!
=1/479,001,600=0.0000000002**0876756987868098979210090321201432312543423654 5 347656458**    7675698786809897921009032    P 54    244=>10=>1

**1/13!**
=1/6,227,020,800=0.00000000001**6059043836821614599392377170154947932725710 5 0348828126**6059043836821614599392377170154947 9    P 54    241=>7

1/14!
=0.0000000000011**470745597729724713851697978682105666232650359634486618 61
3602740586867570994555121539248523375507502491629475756459883444010428 1
3741226439639138051836464534877233289931702630115328528026940725353423 7
6612217882059151900421741691582961424231265501106770948040789310630580 4
7185031312015438999565983692967819951946936073920200904327888454872581 8
5670884083582496280908979321677734376**14707455977297247138516979 7
    P

1/15!
=0.000000000000**76471637318198164759011319857880704441551002397563244124
0906849372457838066303674769283234891700500166108631717097325562934028 5
4249415095975942536789097635658482219328780175341021901868462715023561 5
8440814525470610126694782779438864094948751033407118063202719287375372 0
3145668754134362599971065579531187996796462404928013393621859230324838 7
9044725605572166418727265288111848958409804970651531498092344653191214 0
3777488433573089657745742401827057911713996370081026165682250338334994 4

1965050430658896267361875827484293092758701224309689918155526621135086743552352017960483568949177414785880394346002811611277219742828208436674045139653605262070870536479002087467695933043989128645213301297957382613467269551925636581721237805893890549975206059862144518229174313830398486483142567798652454737110821766906422991079075735160391245047329703414359499015583671668327752983837639922296006952091608176264260920345576430232514888599544684200768856853512938169022825107481192137276793361449446105530761615417700073784729869385954042038698123354208010292666377322461978546634631290715946800602885258969915054571139227223883308539393195477851562507647163731819816475901131985788070444155100239756324412409

                                                              P

**Look at the product of any number with the fractions. Looks as if the pattern still repeats….unbelievable…..**

**[Check this out! 11/83    SDQ(N) = SDQ(D) = 2]**
**Eratosthenes**
**11/83=0.132530120481927710843373493975903614457831325301204819277108433**73493975903614

                                                        P 41     171=>9


**1/83=0.0120481927710843373493975903614457831325**30120

                                                        P 41     **171=>9**


**PATTERNS**
**Postulate: Every prime number ≥ 7 has an inverse whose RB (Repeat Block) SDQ sum is 9.**
**Postulate: No prime number (other than 3) has an SDQ of 3, 6 or 9.**


**Decimal Types**
**A. Non repeating digits**
        1) Finite number of digits, such as 1/8=0.125
        2) Infinite number of digits, such as e, $\pi$ and all irrational numbers, $2^{1/2}$ =
1.414…etc.
**B. Repeating single digit, such as all 3's, as in 1/3=0.333333333333**
The possible single repeating digits are all: 1, 2, 3, 5, 6, or 7.
        All 1's: 1/9, 1/90, 1/900, …
        All 2's: 1/45,
        All 3's: 1/3, 1/12, 1/30, 1/48, 1/75, 1/120, …    SDQ of each denominator = 3

All 4's: **NONE**

All 5's: 1/18,

All 6's: 1/6, 1/15, 1/24, 1/60, 1/96, …          SDQ of each denominator = 6

All 7's: 1/36,

All 8's: **NONE**

## C. Repeat Block RB

1) Repeat Block sum whose single digit quality, SDQ is: 1, 3, 6, or 7

SDQ = 1: 1/27, 1/81, …

SDQ = 3: 1/33, 1/102, 1/123, 1/126, …

SDQ = 6: 1/51, 1/63, 1/66, 1/69, 1/87, 1/117, …

SDQ = 7: 1/108,

2) Repeat Block sum whose single digit quality, SDQ is 9. **Prime denominators are shown in bold type.**

1/7, **1/11**, 1/12, 1/14, **1/17**, **1/19**, 1/21, 1/22, **1/23**, 1/26, 1/28, **1/29**, **1/31**, 1/34,1/35, **1/37**, 1/38, 1/39, **1/41**, 1/42, **1/43**, 1/44, 1/46, **1/47**, 1/49, 1/52, **1/53**, 1/55, 1/56, 1/57, 1/58, **1/59**, **1/61**, 1/62, 1/65, **1/67**, 1/68, 1/70, **1/71**, 1/73, 1/74, 1/76, 1/77, 1/78, **1/79**, 1/82, **1/83**, 1/84, 1/85, 1/86, 1/88, **1/89**, 1/91, 1/92, 1/93, 1/94, 1/95, **1/97**, 1/98, **1/101**, **1/103**, 1/104, 1/105, 1/106, **1/107**, **1/109**, 1/110, 1/111, 1/112, **1/113**, 1/114, 1/115, 1/116, 1/118, 1/119, 1/121, 1/122, 1/124, **1/127**,

**Every prime number ≥ 7 has an inverse whose RB (Repeat Block) SDQ sum is 9.**

**Examples of A. Non repeating digits**

**2) Infinite number of digits, such as π, e and all irrational numbers, √2 = 1.414…etc.**

## ANOTHER PATTERN

Consider the number 3.14159. The first digit is 3. If we add the next digit (1) to 3 we get 4. Adding the next digit (4) to 4 we get 8. Adding the next digit (1) to 8 we get 9. Adding the next digit (5) to 9 we get 14 with a SDQ of 5. Adding the next digit (9) to 5 we get 14 with a SDQ of 5. In this fashion we collect all of these digits to yield, 3.48955. Essentially we are adding all of the preceding digits, always taking their SDQ values and forming a new number.

Let us consider π = 3.141592653589… and use the SDQ to evaluate the sum of the digits.

**3.141592653589**

1) 3.489557493877

2) 3.766275993297

3) 3.174649993553

3.426377771625

3.796975312816

3.117751457674

3.453167275294

3.736742427994

3.141835929994

3.489827799994

3.766575333337

3.174973693697

3.426647447442

3.796375942613

3.117184481782

3.453437212981

3.736142457767


3.141268386418

3.489287196121

3.766875663467

3.174316393742

3.426917114268

3.796675672419

3.117427424899

3.453797248777

3.736442487531

**27) 3.141592653**823


Look for repetition after multiples of 27 iterations.

$2^{1/2}$=1.41421356237309504880168872420969807856967 1875


BE CAREFUL, NOT EXACTLY REPEATABLE. THE PROCESS REMINDS OF
A KIND OF FIBO LIKE SET OF SEQUENCES.
LOOR FOR PATTERNS IN THE COLUMNS OF DIGITS…..


**Examples of C. Repeat Block**
        2) Repeat Block sum whose single digit quality, SDQ is 9. **Prime**

**denominators are shown in bold type.**

## REPEAT BLOCK PATTERNS

We now examine the patterns of the Repeat Blocks, RB, starting with P 2. Prime denominators are indicated in **bold type**.

**P 2**

**Pattern: 1- The inverse of every multiple of the prime 11 yields a P of 2.**
**2- The SDQ of every RB is 9 except when the SDQ of the denominator is 3, 6, or 9.**

1/**11** = 0.**09**0909090909                    P 2      09=>9

| | | |
|---|---|---|
| 1/22 = 0.0**45**454545454 | P 2 | 45=>9 |
| 1/33 = 0.0**30**303030303 | P 2 | 3 |
| 1/44 = 0.02**27**27272727 | P 2 | 9 |
| 1/55 = 0.0**18**181818181 | P 2 | 18=>9 |
| 1/66 = 0.0**15**151515151 | P 2 | 6 |
| 1/**77** = 0.**012987**012987012 | P 6 | 27 => 9 |
| 1/88 = 0.011**36**3636363 | P 2 | 9 |
| 1/**99** = 0.**01**0101010101 | P 2 | 1 |
| 1/110 = 0.00**90**9090909 | P 2 | 9 |
| 1/121 = 0.00**82644628099173553719**00826446281 | | |
| | P 22 | 99 => 18 => 9 |
| 1/132 = 0.00**75**7575757 | P 2 | 12=>3 |
| 1/**143** = 0.00**6993**006993006 | P 6 | 27 => 9 |
| 1/154 = 0.00**649350**6493506 | P 6 | 27 => 9 |
| 1/165 = 0.00**60**6060606 | P 2 | 6 |
| 1/176 = 0.00**568**181818 | P 2 | 9 |
| 1/**187** = 0.00**5347593582887**7005347593582887701 | | |
| | P 16 | 81 => 9 |
| 1/198 = 0.00**50**5050505 | P 2 | 5 |
| 1/209 = 0.00**478468899521531**10047846889952153 | | |
| | P 18 | 81 => 9 |
| 1/220 = 0.00**45**4545454 | P 2 | 9 |

**P 3**

**Pattern: 1- The inverse of 27 and prime 37, excluding multiples of 81, with P 9.**

   **2- The SDQ of the RB is 9 except for denominators with SDQ 9.**

   **3- No 81 or its multiples.**

27, 37, 54, 74, 108, 111, 135, 148, 185,

27, 37, 54, 74, 108, 111, 135, 148, 185,


**No 81 or its multiples**


| | | |
|---|---|---|
| 1/27=0.**037**037037037037… | P 3 | 10=>1 |
| | | |
| **1/37**=0.**027**027027027027027027027027027027 | P 3 | **27=>9** |
| 1/54=0.0**185**185185185185185185185185185 | P 3 | 14=>5 |
| 1/74=0.0**135**135135135135135135135135135 | P 3 | **9** |
| [1/81=0.**012345679**012345679012345679012345 | **P 9** | 37=>10=>1] |
| 1/108=0.009**259**259259259259259259259259259 | P 3 | 16=>7 |
| 1/111=0.009**009**009009009009009009009009009 | P 3 | **9** |
| 1/135=0.007**407**407407407407407407407407407 | P 3 | 11=>2 |
| 1/148=0.006**756**756756756756756756756756756 | P 3 | **18=>9** |
| [1/162=0.006**17283950**617283950617283950617 | **P 9** | 41=>5] |
| 1/185=0.005**405**405405405405405405405054 | P 3 | **9** |


**P 4**

**Pattern: 1- The inverse of every multiple of the prime 101 has a P of 4, except for multiples of 707.**

   **2- Every P of 4 has an SDQ of 9 except for denominators with SDQ of 3, 6 or 9.**


| | | |
|---|---|---|
| 1/**101**=0.**0099**00990099009900 | P 4 | 18=>9 |
| 1/202=0.00**4950**495049504950 | P 4 | 18=>9 |
| 1/303=0.00**3300**330033003300 | P 4 | 6 |
| 1/404=0.002**4752**475247524752475 | P 4 | 18=>9 |
| 1/505=0.00**1980**198019801980 | P 4 | 18=>9 |
| 1/606=0.001**650**165016501650 | P 4 | 12=>3 |
| 1/707=0.001**414427157**001414 | P 12 | 36=>9 |
| 1/808=0.001**237**623762376237 | P 4 | 18=>9 |
| 1/909=0.001**100**110011001100 | P 4 | 2 |
| 1/1010=0.000**99**009900990099 | P 4 | 18=>9 |

**P 5**    Pattern not fully established yet.
Pattern: 1- Inverse of multiples of prime 41 have P 5, except for multiples of 287 having P 30.
2- All RB have SDQ of 9 except when SDQ D=3, 6, 9.


41, 82, 123, 164, 205, 246, <u>287</u>, 328, 369, 410, <u>451</u>, 492, <u>533</u>, <u>574</u>, 615, 656, <u>697</u>, 738, <u>779</u>, 820, 861, <u>902</u>, 943, 984, 1025, <u>1066</u>, 1107, <u>1148</u>, <u>1189</u>,


| | | |
|---|---|---|
| 1/41=0.024390243902439024390243902439024 | P 5 | 18=>9 |
| 1/82=0.012195121951219512195121951219512 | P 5 | 18=>9 |
| 1/123=0.008130081300813 | P 5 | 12=>3 |
| 1/164=0.0060975609756097560975609756097560975609 | P 5 | 27=>9 |
| 1/205=0.00487804878048780487804878 | P 5 | 27=>9 |
| 1/246=0.00406504065040650406504065 | P 5 | 15=>6 |
| 1/<u>287</u>=0.0034843205574912891986062717 7700348 | P 30 | 135=>9 |
| 1/328=0.0030487804878048780487804878048 | P 5 | 27=>9 |
| 1/369=0.002710027100271002710027100 | P 5 | 10=>1 |
| 1/410=0.0024390243902439024390243 90 | P 5 | 18=>9 |
| 1/<u>451</u>=0.0022172949002217294900221 72 | P 10 | 68=>14=>5 |
| 1/492=0.00203252032520325203252032 5 | P 5 | 12=>3 |
| 1/<u>533</u>=0.00187617260787992495309568 480300187 | P 30 | |
| 1/<u>574</u>=0.0017421602787456445993031 35888501742 | P 30 | 135=>9 |
| 1/615= 0.00162601626016260162601626 01626 | P 5 | 15=>6 |
| 1/656=0.00152439024390243902439024 3 | P 5 | 18=>9 |

1/<u>697</u>=0.00143472022955523672883787661406025824964131994261119081779053084648493543758967001434720229555236728 83     P 80
1/738=0.001355013550135501355013550            P 5    14=>5
1/<u>779</u>=0.001283697047496790757381258023106546854942233632862644415917843388960205391527599486521181001283697047 49679

P 90
| 1/820=0.0012195121951219512195121951219512 | P 5 | 18=>9 |
|---|---|---|
| 1/<u>861</u>=0.0011614401858304297328687572 59001161 | P 30 | 126=>9 |
| 1/<u>902</u>=0.001108647450110864745011086 474501108 | P 10 | 36=>9 |


1/<u>943</u>=0.00106044538706256627783669141039236479321314952279957582184517497348886532343584305408271474019088016967126193001060     P 112

**1/984=0.00101626**0162601626016260162601                    P 5      15=>6


**1/1025=0.00097560**975609756097560975609                    P 5      27=>9

**1/1066=0.00093808630393996247654784240150**0938             P 30

1/1107=


1/**1148**=0.000**87108013937282299651567944250**871            P 30

1/**1189**=0.000**84104289318755256518082422203532380151387720773759461732548359966358284272497897392767031118587047939444911690496215306980656013456686291**0008410428931875525651808242220353238015138772077375

                                                               P


1/1394=0.000**717360114777618364418938307030129124820659971305595408895265423242467718794835**0071736011477761    P 80?


Looks like 7x41, 11x41, 13x41, 17x41, 19x41, 23x41, etc any p>5 times 41 and their multiples yields a P other than 5. [check]


**P 6      WOW!**
**Pattern: 1-The inverse of every multiple of the primes 7 and 13 yields a P of 6.**
**          2- The SDQ of every RB is 9, except when the denominator has an SDQ of 9.**

1/7 = 0.**142857**142857                              P 6     **27=>9**
1/**13** = 0.**076923**076923                            P 6     **27=>9**
1/14 = 0.0**714285**714285                            P 6     27=>9
1/21= 0.**047619**047619                              P 6     27=>9
1/26= 0.0**384615**384615…                            P 6     27=>9
1/28= 0.03**571428**571428                            P 6     27=>9
1/35= 0.02**857142**857142                            P 6     27=>9


1/39= 0.02**564102**564102      P 6     18=>9
1/42= 0.02**380952**380952      P 6     27=>9
1/52= 0.019**23076**923076      P 6     27=>9
1/56= 0.017**85714**285714      P 6     27=>9
1/63= 0.**015873**01587301      P 6     24=>6
1/65= 0.01**538461**538461      P 6     27=>9
1/70= 0.01**428571**428571      P 6     27=>9
1/77= 0.0**12987**01298701      P 6     27=>9

| | | |
|---|---|---|
| 1/84= 0.01**190476**190476 | P 6 | 27=>9 |
| 1/91= 0.**01098901098901** | P 6 | 27=>9 |
| 1/104=0.00**96153846**153 | P 6 | 27=>9 |
| 1/112=0.0089**285714**285 | P 6 | 27=>9 |
| 1/117=0.**0085470085470** | P 6 | 24=>6 |
| 1/126=0.00**79365079365** | P 6 | 30=>3 |
| 1/130=0.00**76923076923** | P 6 | 27=>9 |
| 1/140=0.0071**428571**428 | P 6 | 27=>9 |
| 1/143=0.**0069930069930** | P 6 | 27=>9 |
| 1/154=0.00**64935064935** | P 6 | 27=>9 |
| 1/156=0.00**64102564102** | P 6 | 18=>9 |
| 1/168=0.0059**5238095**23 | P 6 | 27=>9 |
| 1/175=0.00**57142857142** | P 6 | 27=>9 |

## REPEAT BLOCK 6 - ANALYSES

The Repeat Block, P of **six digits** in 1/7, namely, **142857** does not contain 0, 3, 6, or 9, and adds to 27=>9.

**SDQ P(1/7) = 9.**

The repeating sequence of **six digits** in 1/13, namely, **076923** does not contain 1, 4, 5, and 8, and they also add to 27=>9.

**SDQ P(1/13) = 9.**

The repeating sequence of **six digits** in 1/14, namely, **714285** does not contain 0, 3, 6, or 9, and they also add to 27=>9.

**SDQ P(1/14) = 9.**

The repeating sequence of sixteen digits in 1/17, namely, **0.588235294117647** contains one each of 0, 3, 6, and 9, and doubles of 1, 2, 4, 5, 7, and 8. Since 1, 2, 4, 5, 7 and 8 add to 27, then double that is 54, and since 0, 3, 6, and 9 add to 18, the total sum is 72=>9.

**SDQ P(1/17) = 9.**

The repeating sequence of 18 digits in 1/19, namely, **052631578947368421** contains one each of 0, and 9 and doubles of 1, 2, 3, 4, 5, 6, 7, and 8. Since 1, 2, 3, 4, 5, 6, 7, and 8 add to 36 and double that is 72, and added to the single 9 is the total sum 81=> 9.

**SDQ P(1/19) = 9.**

The repeating sequence of 6 digits in 1/21, namely, **047619** does not contain 2, 3, 5, or 8, and they also add to 27=>9.

**SDQ P(1/21) = 9.**

Whenever the sum of a sequence has an SDQ of 9, multiplication by any integer will maintain that SDQ of 9. Thus, 2/7 = 2 x (1/7) = 0.**285714**285714... whose sequence is 2,8,5,7,1,4 adding to 27=>9. 3/7 = 3 x (1/7) = 0.**428571**428571... Again this is a six digit sequence adding to 27=>9.

Also notice the digit verse and inverse relationship between 27 and 72.

The unusual qualities of 1/7 will be examined in later in terms of musical and biblical connotations.

**SO FAR NO P 7**

**P 8**

73, 137, 146

**P 9**

81, 112, 113, 162

**P 13**

53, 158, 159, 79, 106

**P 15**

31, 62, 93, 155

**P 16**

17, 34, 51, 68, 85, 102, 153, 170

**P 18**

19, 38, 57, 76, 95, 152, 171

**P 21**

43, 172, 86

**P 22**

23, 46, 69, 92

**P 28**

29, 58, 87, 174

**P 32**

67

**P 34**

103

**P 35**

71

**P 41**

83, 166

**P 42**

49, 98

**P 43**

173
**P 44**
89, 163, 178
**P 46**
47, 94
**P 53**
107
**P 58**
59, 177
**P 60**
61
**P 66**
161
**P 75**
151
**P 78**
157, 169

**P 81**
163
**P 96**
97
**P 112**
113, 162
**P 166**
167

## SDQ AND THE INVERSE OF FACTORIALS

1/1!=1
½!=.5

| | |
|---|---|
| **1/3!=1/6=0.16666666666666666** | all 6's |
| ¼!=1/24=0.041666666666666666 | all 6's |
| **1/5!=1/120=0.00833333333333333** | all 3's |
| 1/6!=1/720=0.00138888888888888 | all 8's |

| | | |
|---|---|---|
| **1/7!=1/5,040=0.00019841269841269** | P 6 | 30=>3 |
| 1/8!=1/40,320=0.000024801587301587 | P 6 | 24=>6 |

1/9!=1/362,880=0.00000027**557319223985890652**55731                 P 18    89=>17=>8

1/10!=1/3,628,800=0.000000027**557319223985890652**55731             P 18    89=>17=>8

**1/11**!=1/39,916,800=0.00000002**5052108385441718775**052           P 18    76=>13=>4

1/12!

=1/479,001,600=0.0000000002**0876756987868098979210090321201432312543423654**5

**347656458**     7675698786809897921009032                          P 54

244=>10=>1


**1/13!**

=1/6,227,020,800=0.00000000001**6059043836821614599392377170154947932725710**5

**0348828126**6059043836821614599392377170154947               9         P 54

241=>7


1/14!

=1/87,178,291,200=0.0000000000011**470745597729724713851697978682105666232**65

**0359634486618613602740586867570994555121539248523375507502491629475756**4

**5988344401042813741226439639138051836464534877233289931702630115328528**0

**2694072535342376612217882059151900421741691582961424231265501106770948**0

**4078931063058047185031312015438999565983692967819951946936073920200904**3

**2788845487258185670884083582496280908979321677734376**1470745597729724713

8516979786821056662                                       P


1/15!

=1,307,674,368,000=0.000000000000**0764716373181981647590113198578807044415**5

**1002397563244124090684937245783806630367476928323489170050016610863171**7

**0973255629340285424941509597594253678909763565848221932878017534102190**1

**8684627150235615844081452547061012669478277943886409494875103340711806**3

**2027192873753720314566875413436259997106557953118799679646240492801339**3

**6218592303248387904472560557216641872726528811184895840980497065153149**8

**0923446531912140377748843357308965774574240182705791171399637008102616**5

**6822503383349944196505043065889626736187582748429309275870122430968991**8

**1555266211350867435523520179604835689491774147858803943460028116112772**1

**9742828208436674045139653605262070870536479002087467695933304398912864**5

**2133012979573826134672695519256365817212378058938905499752060598621445**1

**8229174313830398486483142567798652454737110821766906422991079075735160**3

**9124504732970341435949901558367166832775298383763992229600695209160817**6

**2642609203455764302325148885995446842007688568535129381690228251074811**9

**2137276793361449446105530761615417700073784729869385954042038698123354**2

**0801029266637732246197854663463129071594680060288525896991505457113 9227**
**22388330853939319547785156250**7647163731819816475901131985788070444155 10
0239756324412409068                          P

**P 7**

**P 8**
**Look at the product of any number with the fractions. Looks as if the pattern still repeats....unbelievable.....**
**Check this out! 11/83**
**Eratosthenes**
**11/83=0.13253012048192771084337349397590361445783**13253012048192771084337
3493975903614                                         P 41      171=>9

67

# Chapter 4
## Single Digit Quality: Exponentials

We now examine the single digit quality applied to exponentials, numbers that are multiples of themselves. Exponentials are formed from two numbers, say x and n. The number x is called the base and n is the exponent of that number. This is shorthand and means that x is multiplied by itself n times. This is written mathematically as $x^n$. For example, if the number that we want is 2 multiplied by itself three times, we say that the number 2 is raised to the exponent 3. This equals 2 times 2 times 2 = 8 or $2^3 = 8$. Again, $x^n$ means that the value x is raised to the power n, thus $2^3$ means 2 raised to the 3 power, which means 2x2x2=8. It is not necessary that either x or n be integers. Table 4-1 lists integer values of x and n, and their values when exponentiated. The left-hand most column lists x and the top row across the top lists n.

**TABLE 4-1: EXPONENTIAL VALUES**

| n \ x | 0 | 1 | 2 | 3 | 4 | 5 | 6 |
|---|---|---|---|---|---|---|---|
| 1 | 1 | 1 | 1 | 1 | 1 | 1 | 1 |
| 2 | 1 | 2 | 4 | 8 | 16 | 32 | 64 |
| 3 | 1 | 3 | 9 | 27 | 81 | 243 | 729 |
| 4 | 1 | 4 | 16 | 64 | 256 | 1024 | 4096 |
| 5 | 1 | 5 | 25 | 125 | 625 | 3125 | 15625 |
| 6 | 1 | 6 | 36 | 216 | 1296 | 7776 | 46656 |
| 7 | 1 | 7 | 49 | 343 | 2401 | 16807 | 117649 |
| 8 | 1 | 8 | 64 | 512 | 4096 | 32768 | 262144 |
| 9 | 1 | 9 | 81 | 729 | 6561 | 59049 | 531441 |
| 10 | 1 | 10 | 100 | 1000 | 10000 | 100000 | 1000000 |
| 11 | 1 | 11 | 121 | 1331 | 14641 | 161051 | 1771561 |
| 12 | 1 | 12 | 144 | 1728 | 20736 | 248832 | 2985984 |
| 13 | 1 | 13 | 169 | 2197 | 28561 | 371293 | 4826809 |
| 14 | 1 | 14 | 196 | 2744 | 38416 | 537824 | 7529536 |
| 15 | 1 | 15 | 225 | 3375 | 50625 | 759375 | 11390625 |
| 16 | 1 | 16 | 256 | 4096 | 65536 | 1048576 | 16777216 |
| 17 | 1 | 17 | 289 | 4913 | 83521 | 1419857 | 24137569 |
| 18 | 1 | 18 | 324 | 5832 | 104976 | 1889568 | 34012224 |

| 19 | 1 | 19 | 361 | 6859 | 130321 | 2476099 | 47045881 |
|----|---|----|-----|------|--------|---------|----------|
| 20 | 1 | 20 | 400 | 8000 | 160000 | 3200000 | 64000000 |
| 21 | 1 | 21 | 441 | 9261 | 194481 | 4084101 | 85766121 |
| 22 | 1 | 22 | 484 | 10648 | 234256 | 5153632 | 113379904 |

In Table 4-2, we see the same table showing the SDQ for all values with the exponent n extended to 12. For example; in the last row, last column, SDQ($22^6$) = SDQ(113379904) = 1+1+3+3+7+9+9+0+4 = 37 = 3+7 = 10 = 1+0 = 1. Thus, the SDQ of the number $22^6$=113379904 is 1. In mathematical terms these are exponents are in mod 9.

TABLE 4-2: TABLE OF EXPONENTIAL QUALITIES (SDQ)

| n | 0 | 1 | | 2 | 3 | 4 | 5 | 6 | 7 | | 8 | 9 | 10 | 11 | 12 | 13 | | 14 |
|---|---|---|---|---|---|---|---|---|---|---|---|---|----|----|----|----|---|----|
| x | | | | | | | | | | | | | | | | | | |
| 1 | 1 | 1 | | 1 | 1 | 1 | 1 | 1 | 1 | | 1 | 1 | 1 | 1 | 1 | 1 | | 1 |
| 2 | 1 | 2 | | 4 | 8 | 7 | 5 | 1 | 2 | | 4 | 8 | 7 | 5 | 1 | 2 | | 4 |
| 3 | 1 | 3 | | 9 | 9 | 9 | 9 | 9 | 9 | | 9 | 9 | 9 | 9 | 9 | 9 | | 9 |
| 4 | 1 | 4 | | 7 | 1 | 4 | 7 | 1 | 4 | | 7 | 1 | 4 | 7 | 1 | 4 | | 7 |
| 5 | 1 | 5 | | 7 | 8 | 4 | 2 | 1 | 5 | | 7 | 8 | 4 | 2 | 1 | 5 | | 7 |
| 6 | 1 | 6 | | 9 | 9 | 9 | 9 | 9 | 9 | | 9 | 9 | 9 | 9 | 9 | 9 | | 9 |
| 7 | 1 | 7 | | 4 | 1 | 7 | 4 | 1 | 7 | | 4 | 1 | 7 | 4 | 1 | 7 | | 4 |
| 8 | 1 | 8 | | 1 | 8 | 1 | 8 | 1 | 8 | | 1 | 8 | 1 | 8 | 1 | 8 | | 1 |
| 9 | 1 | 9 | | 9 | 9 | 9 | 9 | 9 | 9 | | 9 | 9 | 9 | 9 | 9 | 9 | | 9 |
| | | | | | | | | | | | | | | | | | | |
| 10 | 1 | 1 | | 1 | 1 | 1 | 1 | 1 | 1 | | 1 | 1 | 1 | 1 | 1 | 1 | | 1 |
| 11 | 1 | 2 | | 4 | 8 | 7 | 5 | 1 | 2 | | 4 | 8 | 7 | 5 | 1 | 2 | | 4 |
| 12 | 1 | 3 | | 9 | 9 | 9 | 9 | 9 | 9 | | 9 | 9 | 9 | 9 | 9 | 9 | | 9 |
| 13 | 1 | 4 | | 7 | 1 | 4 | 7 | 1 | 4 | | 7 | 1 | 4 | 7 | 1 | 4 | | 7 |
| 14 | 1 | 5 | | 7 | 8 | 4 | 2 | 1 | 5 | | 7 | 8 | 4 | 2 | 1 | 5 | | 7 |
| 15 | 1 | 6 | | 9 | 9 | 9 | 9 | 9 | 9 | | 9 | 9 | 9 | 9 | 9 | 9 | | 9 |
| 16 | 1 | 7 | | 4 | 1 | 7 | 4 | 1 | 7 | | 4 | 1 | 7 | 4 | 1 | 7 | | 4 |
| 17 | 1 | 8 | | 1 | 8 | 1 | 8 | 1 | 8 | | 1 | 8 | 1 | 8 | 1 | 8 | | 1 |

| 18 | 1 | 9 |  | 9 | 9 | 9 | 9 | 9 | 9 |  | 9 | 9 | 9 | 9 | 9 | 9 |  | 9 |
|----|---|---|--|---|---|---|---|---|---|--|---|---|---|---|---|---|--|---|
|    |   |   |  |   |   |   |   |   |   |  |   |   |   |   |   |   |  |   |
| 19 | 1 | 1 |  | 1 | 1 | 1 | 1 | 1 | 1 |  | 1 | 1 | 1 | 1 | 1 | 1 |  | 1 |
| 20 | 1 | 2 |  | 4 | 8 | 7 | 5 | 1 | 2 |  | 4 | 8 | 7 | 5 | 1 | 2 |  | 4 |
| 21 | 1 | 3 |  | 9 | 9 | 9 | 9 | 9 | 9 |  | 9 | 9 | 9 | 9 | 9 | 9 |  | 9 |
| 22 | 1 | 4 |  | 7 | 1 | 4 | 7 | 1 | 4 |  | 7 | 1 | 4 | 7 | 1 | 4 |  | 7 |
|    |   |   |  |   |   |   |   |   |   |  |   |   |   |   |   |   |  |   |

Notice the regularity of the SDQ numbers. When the exponent $n = 0$, all of the values of $x^0$ are 1, and when $n = 1$, all of the values of $x^1$ are just x. Other than $n = 0$ and 1, there is a regular pattern for both n and x. We only require the six n values from $n = 2$ to $n = 7$, and the nine x values from $x = 1$, to $x = 9$, in order to determine every other x and n SDQ value. This is indicated in the table by the empty columns and rows showing that all of the values are contained in one subtable. Again, each of the base x values, from 1 to 9, produces a periodic sequence of values when raised to the powers represented by the values of n. Thus, for a base value of $x=2$, the SDQ values from $n = 0$ to $n= 12$, are respectively, 1 2 4 8 7 5 1 2 4 8 7 5 . . .  Also, there is the remarkable feature that when the base x becomes equal to 10, 19, etc, the overall $x=1$ to $x=9$ pattern repeats, thus, even the x values can be reduced to their SDQ and continue to yield the same SDQ values for the various powers. Going back to our previous example; $SDQ(22^6) = SDQ(4^6) =1$. To find the SDQ of any base x raised to any power n, we need only the values of x from 2 to 9, and the values of n from 0 to 5.

**The SDQ results of any base x are the same as the SDQ(x).  A remarkable feature evident in the table, is that other than $n = 1$, there is never 3 or 6 as a resultant SDQ.**

## SUMMARY:  BASE AS PARAMETER AND VARIABLE EXPONENT
Let us examine the patterns of the ten sequence terms:
- The $0^0$ sequence terms obviously produces all zeros, as expected.
- The $1^n$ sequence terms produces all ones, as expected.
- The $2^n$ terms produce the repeating sequence **1 2 4 8 7 5**, **1 2 4 8 7 5**, and so on. This sequence contains six qualities, three odd, **1,7,5** and three even, **2,4,8**. Notice that $11^n$, and $20^n$ produce the same sequence terms as $2^n$, since $SDQ(11) = SDQ(20) = 2$.

- The $3^n$ terms produce the sequence **1 3 9 9 9 ...**

- The $4^n$ terms produce the repeating sequence **1 4 7** ... The sum of the sequence terms is **12 = SDQ(3).**

- The $5^n$ terms produce **1 5 7 8 4 2, 1 5 7 8 4 2,** which is like the $2^n$ terms in that they both start with 1, with the remaining terms of the $5^n$ sequence being the reverse of the $2^n$ sequence terms. Thus $5^n$ and $2^n$ are of similar form. The sum of the sequence terms of both the $2^n$ and the $5^n$, are is **27 = SDQ(9).**

- The $6^n$ terms produce **1 6 9 9 9 ....** This is similar to the $3^n$ sequence except that here the **3** term is here replaced by a 6.

- The $7^n$ terms produce **1 7 4 ...** as the repeating sequence, similar to the $4^n$ sequence, with the **4 7** terms reversed. As in the $4^\wedge n$ sequence, the terms sum to **12 = SDQ(3).**

- The $8^n$ terms produce **1 8 ...** as the repeating sequence, which adds to **9.**

- The $9^n$ terms produce **1 9 9 9 ....** as the repeating sequence, which adds to **10 = SDQ(1).**

   If we look at the pattern of terms for the powers of 2, we get; 1 2 4 8 7 5. Ignoring the 1, which appears as the first term in all the x values, excepting zero, we get the pattern 2 4 8 7 5. Notice that patterns for the power of 2 and the power of 5 are similar, as are the patterns for the power of 4 and the power of 7. The patterns for the powers of 3 6 and 9 are also similar: 1 3 9 followed by all nines, 1 6 9 followed by all nines, and 1 9 9 followed by all nines, respectively.

   The powers of 8, produces the sequence of terms 1 8 ...

   Later (chapter    ), we show that the x = 2 is particularly interesting for music applications, since each octave doubles the frequency of the previous one. The base octave is related to the 1, the first octave to the 2, the second to the 4, the third to the 8, the fourth to the 7, and the fifth to the 5. The sixth octave repeats the cycle.

   All power sequences follow a simple pattern when reduced to a single digit, and any power or power series with any large numbers and large exponents may be tested for incorrectness by using this simple technique.

   Table 4-3 lists the exponential qualities with the minimum required bases to reproduce all base SDQ values.

## TABLE 4-3: MINIMUM BASE EXPONENTIAL VALUES

| n | 0 | 1 | | 2 | 3 | 4 | 5 | 6 | 7 |
|---|---|---|---|---|---|---|---|---|---|
| **SDQ(X)** | | | | | | | | | |
| 1 | 1 | 1 | | 1 | 1 | 1 | 1 | 1 | 1 |
| 2 | 1 | 2 | | 4 | 8 | 7 | 5 | 1 | 2 |
| 3 | 1 | 3 | | 9 | 9 | 9 | 9 | 9 | 9 |
| 4 | 1 | 4 | | 7 | 1 | 4 | 7 | 1 | 4 |
| 5 | 1 | 5 | | 7 | 8 | 4 | 2 | 1 | 5 |
| 6 | 1 | 6 | | 9 | 9 | 9 | 9 | 9 | 9 |
| 7 | 1 | 7 | | 4 | 1 | 7 | 4 | 1 | 7 |
| 8 | 1 | 8 | | 1 | 8 | 1 | 8 | 1 | 8 |
| 9 | 1 | 9 | | 9 | 9 | 9 | 9 | 9 | 9 |

## USING THE MINIMUM SDQ TABLE

### MINIMUM SDQ EXPONENTIAL TABLE

| n | 0 | 1 | | 2 | 3 | 4 | 5 | 6 | 7 |
|---|---|---|---|---|---|---|---|---|---|
| | | | R= | 0 | 1 | 2 | 3 | 4 | 5 |
| **SDQ(x)** | | | | | | | | | |
| 1 | 1 | 1 | | 1 | 1 | 1 | 1 | 1 | 1 |
| 2 | 1 | 2 | | 4 | 8 | 7 | 5 | 1 | 2 |
| 3 | 1 | 3 | | 9 | 9 | 9 | 9 | 9 | 9 |
| 4 | 1 | 4 | | 7 | 1 | 4 | 7 | 1 | 4 |
| 5 | 1 | 5 | | 7 | 8 | 4 | 2 | 1 | 5 |
| 6 | 1 | 6 | | 9 | 9 | 9 | 9 | 9 | 9 |
| 7 | 1 | 7 | | 4 | 1 | 7 | 4 | 1 | 7 |
| 8 | 1 | 8 | | 1 | 8 | 1 | 8 | 1 | 8 |
| 9 | 1 | 9 | | 9 | 9 | 9 | 9 | 9 | 9 |

**Exponential SDQ Patterns (excluding n = 0, 1)**

1- There is never a **3** or **6**.

2- Every **3**, **6** or **9** base SDQ will yield a **9**.

3- Every **8** base SDQ will yield an **8** or a **1**, successively.

4- Every **4** and **7** base SDQ will yield a **4**, **7** or **1**.

5- Every **2** and **5** base SDQ will yield a **2**, **5**, **1**, **4**, **7** or **8**.


## Using the Minimum SDQ Exponential Table

**Rule: To use the table to calculate the SDQ($x^n$). Take SDQ(x), then find R = Remainder of (n-2)/6 and locate the cell with the (SDQ(x), R) answer.**

Example: To find the SDQ($25^{14}$), here x = 25 and n = 14. SDQ(25) => 7, and (n-2)/6 = (14-2)/6 = 2 remainder R = 0. Looking at the table we see that row 7, column 0 has the value 4. SDQ($25^{14}$) = 4.

Thus, $25^{14}$ = 37,252,902,984,619,140,625 => 85 => 13 => 4.


Using the Minimum SDQ Exponential Table show that each of the following results are incorrect.

a) $2^{15}$ =? 32,868

b) $16^7$ =? 268,335,456

c) $8^{23}$ =? 590,295,810,358,305,651,712

d) $147^{28}$ =?
4,840,445,926,998,527,143,180,132,566,802,461,408,607,116,960,093,883,732,904,561

a) Here x = 2, and n = 15. (15-2)/6 = 2 R 1, so that SDQ(2) = 2, and R = 1. From the table the answer is 8. Thus, SDQ($2^{15}$) = 8, and since SDQ(32,868) = 27 => 9 this answer must be incorrect.

Correct answer: (32,768) => 26 => 8.


b) $16^7$ =? 268,335,456.

SDQ(16) = 7. For n = 7 we need not use the remainder calculation. From the table the answer is 7. SDQ(268,335,456) = 42 => 6. So this answer is incorrect.

Correct answer: (268,435,456) => 43 => 7.


c) $8^{23}$ =? 590,295,810,358,305,651,712 => 85 => 13 => 4.

 (23 -- 2)/6 = 3 R = 3

From the table we get 8. So this answer is incorrect.

 Correct answer: (590,295,810,358,705,651,712) => 89 => 17 => 8.

**d) $147^{28} =$?**

4,840,445,926,998,527,143,180,132,566,802,461,408,607,116,960,093,883,732,904,561

=> 263 => 11 => 2

SDQ(147) = 12 => 3

(28-2)/6 = 4 R 2

From the table the answer is 9. Therefore the answer given is incorrect.

Correct answer:

(4,840,445,926,998,527,143,180,132,566,802,461,408,607,116,960,091,883,732,904,56
1) => 261 => 9.

## EXAMPLES USING THE SDQ

SDQ(x=2):

Consider any value for x, say x = 2. To find the SDQ for x raised to the nth power, we notice that there are six possibilities for the result, and in the following order; 1 2 4 8 7 and 5, whereupon this sequence repeats itself. We need only divide the value of the exponent, n, by 6, and look at the remainder. If the remainder is 0, then the SDQ is 1; if the remainder is 1, then the SDQ is 2; if the remainder is 2, then the SDQ is 4; remainder of 3, SDQ is 8; remainder of 4, SDQ is 7; remainder of 5, SDQ is 5.

Thus, if $x = 2^{36}$, then n = 36. We divide 36 by 6, and we get as the answer 6 with a remainder of 0, thus the SDQ should be 1. Indeed, $2^{36} = 68,719,476,736 ==> 64 ==> 10 ==> 1$.

| REMAINDER | SDQ VALUE |
|-----------|-----------|
| 0 | 1 |
| 1 | 2 |
| 2 | 4 |
| 3 | 8 |
| 4 | 7 |
| 5 | 5 |

Suppose x=47:

We notice that x = 47 = 11 = 2, so that this value for x yields the same result as x = 2. we want to check SDQ of $47^5$. The result is the same as the SDQ of $2^5$, which is the same as a remainder of 5, and thus reduces to a 5. Indeed, the SDQ of $47^5 = 229,345,007 ==> 32 ==> 5$.

x=3:

Here we note that the first term is 1, the second, 3, and all 9's thereafter. Thus the SDQ of $3^{19}$ = 9. Indeed, $3^{19}$ = 1,162,261,467 ==> 36 ==> 9.

x=3:

Here we note that the first term is 1, the second, 3, and all 9's thereafter. Thus the SDQ of 319 = 9. Indeed, 319 = 1,162,261,467 ==> 36 ==> 9.

Suppose x=21:

We notice that x = 21 = 3, so we are back to same result as for x = 3. Consider $21^7$, we should get 9 as the result. Indeed, we see that $21^7$ = 1,801,088,541 ==> 36 ==> 9.

x=4:

Here, the SDQ follows the repetitive pattern; 1 4 7, and has a repeat cycle of three. We need to divide by 3 and use the remainder of 0, 1, or 2. If the remainder is 0, then the SDQ equals 1; is the remainder is 1, then the SDQ equals 4; and if the remainder is 2, the SDQ equals 7. Thus the SDQ of $31^5$, is the same as the SDQ of $4^5$, which has a remainder of 2, since 5/3 = 1 r 2. Remainder 2, produces a 7. Indeed, the SDQ of $31^5$ = SDQ of 28,629,151 = 2+8+6+2+9+1+5+1 ==> 34 ==> 7.

| REMAINDER | SDQ VALUE |
|:---:|:---:|
| 0 | 1 |
| 1 | 4 |
| 2 | 7 |

x=5:

This case is like x = 2, where we also divide the exponent by 6, but use the following values for the remainders;

| REMAINDER | SDQ VALUE |
|:---:|:---:|
| 0 | 1 |
| 1 | 5 |
| 2 | 7 |
| 3 | 8 |
| 4 | 4 |
| 5 | 2 |

X=6:

The SDQ of powers of 6, are 1 6 9 9 . . .. Thus, any x=6, or any SDQ of x that equals 6, yields the same result, namely, 1 6 9 followed by all 9's. Example: $6^{11}$ = 362,797,056, and therefore, $6^{11}$ ==> 3+6+2+7+9+7+0+5+6 ==> 45 ==> 9, as expected.

Likewise, suppose we consider the example $42^5$. We expect that the SDQ of $42^5$ should equal 9, and indeed, $42^5$ ==> 130,691,232 ==> 1+3+0+6+9+1+2+3+2 ==> 27 ==> 9.

x=7:

The SDQ of powers of 7 yields the repetitive sequence; 1 7 4 with a cycle period of three. Thus we divide the exponent of all base 7, or SDQ equal to 7, by 3 and use the following remainder-SDQ values:

| REMAINDER | SDQ VALUE |
|-----------|-----------|
| 0 | 1 |
| 1 | 7 |
| 2 | 4 |

x=8:

The SDQ of powers of 8 yields the repetitive sequence; 1 8 with a cycle period of two. Thus we divide the exponent of all base 8, or SDQ equal to 8, by 2 and use the following remainder-SDQ values:

| REMAINDER | SDQ VALUE |
|-----------|-----------|
| 0 | 1 |
| 1 | 8 |

X=9:

Here, we get 1 and then all 9's.

The following Table 4-4 is another way of exhibiting the SDQ patterns of any base number and power. The first row is the base number SDQ, followed by rows corresponding to the SDQ values of the powers of that base number.

TABLE 4-4

| Number | 1 | 2 | 3 | 4 | 5 | 6 | 7 | 8 | 9 |
|---|---|---|---|---|---|---|---|---|---|
| SDQ of Square | 1 | 4 | 9 | 7 | 7 | 9 | 4 | 1 | 9 |
| SDQ of Cube | 1 | 8 | 9 | 1 | 8 | 9 | 1 | 8 | 9 |
| SDQ of Fourth | 1 | 7 | 9 | 4 | 4 | 9 | 7 | 1 | 9 |
| SDQ of Fifth | 1 | 5 | 9 | 7 | 2 | 9 | 4 | 8 | 9 |
| SDQ of Sixth | 1 | 1 | 9 | 1 | 1 | 9 | 1 | 1 | 9 |
| SDQ of Seventh | 1 | 2 | 9 | 4 | 5 | 9 | 7 | 8 | 9 |
| SDQ of Eighth | 1 | 4 | 9 | 7 | 7 | 9 | 4 | 1 | 9 |
| SDQ of Ninth | 1 | 8 | 9 | 1 | 8 | 9 | 1 | 8 | 9 |
| SDQ of Tenth | 1 | 7 | 9 | 4 | 4 | 9 | 7 | 1 | 9 |
| SDQ of 11th | 1 | 5 | 9 | 7 | 2 | 9 | 4 | 8 | 9 |
| SDQ of 12th | 1 | 1 | 9 | 1 | 1 | 9 | 1 | 1 | 9 |
| SDQ of 13th | 1 | 2 | 9 | 4 | 5 | 9 | 7 | 8 | 9 |
| SDQ of 14th | 1 | 4 | 9 | 7 | 7 | 9 | 4 | 1 | 9 |
| SDQ of 15th | 1 | 8 | 9 | 1 | 8 | 9 | 1 | 8 | 9 |
| SDQ of 16th | 1 | 7 | 9 | 4 | 4 | 9 | 7 | 1 | 9 |
| SDQ 17th | 1 | 5 | 9 | 7 | 2 | 9 | 4 | 8 | 9 |
| SDQ 18th | 1 | 1 | 9 | 1 | 1 | 9 | 1 | 1 | 9 |
| SDQ 19th | 1 | 2 | 9 | 4 | 5 | 9 | 7 | 8 | 9 |
| SDQ 20th | 1 | 4 | 9 | 7 | 7 | 9 | 4 | 1 | 9 |
| SDQ 21st | 1 | 8 | 9 | 1 | 8 | 9 | 1 | 8 | 9 |
| SDQ 22nd | 1 | 7 | 9 | 4 | 4 | 9 | 7 | 1 | 9 |

Example: Consider the number $47^{15}$ = 12,063,348,350,820,368,238,715,343.

We can use the table to verify if this answer is wrong. We need only look at the SDQ values to instantly make that determination.

SDQ(47) = 2    (since 4+7=11 and 1+1=2) and we therefore want the 2 column.

The exponent 15 when divided by 65 has a remainder of 3, so we are looking for the cube row.

Looking at the table, we see that we are dealing with $2^{15}$

We go to the number 2 in the first row and then down to the row with the SDQ of 15 and find the digit 8.

Therefore the SDQ($47^{15}$) = 8 and if the SDQ of 12,063,348,350,820,368,238,715,343 is NOT 8, then it is the wrong value.

SDQ(12,063,348,350,820,368,238,715,343) = 98=8 and we can say that this result is not wrong. We cannot say if it is correct since interchanging any two digits will produce the same SDQ value.

We see that there are some striking patterns. The SDQ of Square, Eighth power, Fourteenth power, and Twentieth power have the same numbers across. The SDQ of the Cube, Ninth, Fifteenth, powers are the same across. The SDQ of the Fourth, Tenth, Sixteenth, Twenty Second are the same across.

In fact there are only six different pattern sets of numbers across the rows.

1- Square Pattern:  Digits 1, 4, 7, and 9

2- Cubic Pattern: Digits 1, 8 and 9

3- Fourth Power Pattern: Digits 1, 4, 7 and 9

4- Fifth Power Pattern: Digits 1, 2, 4, 5, 7, 8 and 9

5- Sixth Power Pattern: Digits 1 and 9

6- Seventh Power Pattern: Digits 1, 2, 4, 5, 7, 8 and 9

The digits 3 and 6 are missing from every pattern.

## SUMMARY

The remarkable result of the SDQ of exponents is that every integer, raised to the power of any other integer, will yield an SDQ given by the simple results of Table 4.

TABLE 4: REDUCED SDQ TABLE

| n | 0 | 1 | | 2 | 3 | 4 | 5 | 6 | 7 |
|---|---|---|---|---|---|---|---|---|---|
| **SDQ(X)** | | | | | | | | | |
| 1 | 1 | 1 | | 1 | 1 | 1 | 1 | 1 | 1 |
| 2 | 1 | 2 | | 4 | 8 | 7 | 5 | 1 | 2 |
| 3 | 1 | 3 | | 9 | 9 | 9 | 9 | 9 | 9 |
| 4 | 1 | 4 | | 7 | 1 | 4 | 7 | 1 | 4 |
| 5 | 1 | 5 | | 7 | 8 | 4 | 2 | 1 | 5 |
| 6 | 1 | 6 | | 9 | 9 | 9 | 9 | 9 | 9 |
| 7 | 1 | 7 | | 4 | 1 | 7 | 4 | 1 | 7 |
| 8 | 1 | 8 | | 1 | 8 | 1 | 8 | 1 | 8 |
| 9 | 1 | 9 | | 9 | 9 | 9 | 9 | 9 | 9 |

Table 4 lists the minimum required SDQ values for checking calculations for incorrectness. Actually, there are fewer values required for some of the bases, when the number of terms in the sequence is less than 5.

**NEW TOOL FOR TESTING CALCULATIONS**

Therefore, the results of any calculation of exponents, <u>no matter how large the numbers</u>, may be very simply checked using the foregoing SDQ analysis. We note that this check will determine if a calculation is **wrong**, that is, when it does not yield the proper SDQ value. When the SDQ of a calculation yields the correct value, the calculation could still be incorrect, due to interchange of digits, or loss or addition of digits that just happens to produce the correct SDQ. Thus, SDQ analysis is a powerful and simple tool to determine if a calculation produces an incorrect result.

**RESULTS FOR EXPONENT AS PARAMETER AND VARIABLE BASE**

Here we examine Table 3, and look at the pattern of different base values, x, for a particular exponent.

$SDQ(x^2)$ has the values   1,4,9,7,7,9,4,1,9; ...
$SDQ(x^3)$ has the values   1,8,9; ...
$SDQ(x^4)$ has the values  1,7, 9, 4, 4, 9, 7, 1, 9; ...
$SDQ(x^5)$ has the values  1,5,9,7,2,9,4,8,9; . . .
$SDQ(x^6)$ has the values  1,1,9; . . .
$SDQ(x^7)$ has the values  1,2,9,4,5,9,7,8,9; . . .

$SDQ(x^8)$ has the values  1,4,9,7,7,9,4,1,9; . . . $= SDQ(x^2)$
$SDQ(x^9)$ has the values  1,8,9; . . .                    $= SDQ(x^3)$
$SDQ(x^{10})$ has the values 1,7,9,4,4,9,7,1,9; . . . $= SDQ(x^4)$
$SDQ(x^{11})$ has the values 1,5,9,7,2,9,4,8,9; . . . $= SDQ(x^5)$
$SDQ(x^{12})$ has the values 1,1,9; . . .                   $= SDQ(x^6)$
$SDQ(x^{13})$ has the values 1,2,9,4,5,9,7,8,9; . . . $= SDQ(x^7)$

$SDQ(x^{14}) = SDQ(x^8)  = SDQ(x^2)$
$SDQ(x^{15}) = SDQ(x^9)  = SDQ(x^3)$
$SDQ(x^{16}) = SDQ(x^{10}) = SDQ(x^4)$
$SDQ(x^{17}) = SDQ(x^{11}) = SDQ(x^5)$
$SDQ(x^{18}) = SDQ(x^{12}) = SDQ(x^6)$
$SDQ(x^{19}) = SDQ(x^{13}) = SDQ(x^7)$

We observe that the pattern follows the $x^2$ to $x^7$ sequence, by multiples of 6 for the exponents. Thus, to find the equivalent SDQ of say $x^n$; subtract 6 from n, until the number reached is in the range of 2 to 7; then, use that value for the same SDQ.

EXAMPLE 1: If we want the SDQ($x^{47}$), since n=47, here, 7x6 = 42, and 47 - 42 = 5. Thus, SDQ($x^{47}$) = SDQ($x^5$) having the values; 1,5,9,7,2,9,4,8,9; . . ..

EXAMPLE 2: If we want the SDQ($x^{20}$); then since 20 = 6*3 + 2, the result is the same as SDQ($x^2$), with the values; 1,4,9,7,7,9,4,1,9;. Again SDQ($x^{20}$) = SDQ($x^2$).

## ADDITIONAL FEATURES

We may look at these results in another way, by asking the question: which exponent, n, no matter what the fixed base, will yield the same SDQ?

| SDQ | n |
|-----|---|
| 1 | all |
| 2 | 2,5,7 |
| 3 | none |
| 4 | 2,4,5,7 |
| 5 | 2,5,7 |
| 6 | none |
| 7 | 2,4,5,7 |
| 8 | 3,5,7 |
| 9 | all; 2,3,4,5,6,7 |

N.B. We are omitting the obvious results for x1 which are not repetitive, and 9.

Furthermore, there is only one way of producing an SDQ of 3 or an SDQ of 6. Thus, if an SDQ of 3 or 6 is obtained, it can only be due to the fact that 3 to the first power is 3, and 6 to the first power is 6. Notice the relationship between the SDQ pairs of values: 2 and 5, 3 and 6, and 4 and 7. Each of these pairs of SDQ values, whose difference is 3, is generated by the same SDQ(x) values. This regularity of number, is startling, to say the least.

We now have the capability of taking the resulting SDQ of any exponent, and then determining its possible bases from these simple results.

TABLE 4-5 MINIMUM SDQ MULTIPLICATION TABLE

|   | 1 | 2 | 3 | 4 | 5 | 6 | 7 | 8 | 9 |
|---|---|---|---|---|---|---|---|---|---|
| 1 | 1 | 2 | 3 | 4 | 5 | 6 | 7 | 8 | 9 |
| 2 | 2 | 4 | 6 | 8 | 1 | 3 | 5 | 7 | 9 |
| 3 | 3 | 6 | 9 | 3 | 6 | 9 | 3 | 6 | 9 |
| 4 | 4 | 8 | 3 | 7 | 2 | 6 | 1 | 5 | 9 |
| 5 | 5 | 1 | 6 | 2 | 7 | 3 | 8 | 4 | 9 |
| 6 | 6 | 3 | 9 | 6 | 3 | 9 | 6 | 3 | 9 |
| 7 | 7 | 5 | 3 | 1 | 8 | 6 | 4 | 2 | 9 |
| 8 | 8 | 7 | 6 | 5 | 4 | 3 | 2 | 1 | 9 |
| 9 | 9 | 9 | 9 | 9 | 9 | 9 | 9 | 9 | 9 |

1 x 1; 2 x 5; 4 x 7; and 8 x 8. This gives us a clue as to which inverses have SDQ's that are available for analysis.

**SINGLE DIGIT QUALITY: FACTORIALS**

The following table 4-6 illustrates the values of the factorials and their SDQ for the first 22 numbers.

TABLE 4-6: VALUES OF FACTORIALS - SDQ

| n | n! | SUM OF DIGITS | SDQ |
|---|---|---|---|
| 1 | 1 | 1 | 1 |
| 2 | 2 | 2 | 2 |
| 3 | 6 | 6 | 6 |
| 4 | 24 | 6 | 6 |
| 5 | 120 | 3 | 3 |
| 6 | 720 | 9 | 9 |
| 7 | 5,040 | 9 | 9 |
| 8 | 40,320 | 9 | 9 |
| 9 | 362,880 | 27 | 9 |
| 10 | 3,628,800 | 27 | 9 |
| 11 | 39,916,800 | 36 | 9 |
| 12 | 479,001,600 | 27 | 9 |
| 13 | 6,227,020,800 | 27 | 9 |
| 14 | 87,178,291,200 | 45 | 9 |
| 15 | 1,307,674,368,000 | 45 | 9 |
| 16 | 20,922,789,888,000 | 63 | 9 |
| 17 | 355,687,428,096,000 | 63 | 9 |
| 18 | 6,402,373,705,728,000 | 54 | 9 |
| 19 | 121,645,100,408,832,000 | 45 | 9 |
| 20 | 2,432,902,008,176,640,000 | 54 | 9 |
| 21 | 51,090,942,171,709,440,000 | 63 | 9 |
| 22 | 1,124,000,727,777,607,680,000 | **72** | 9 |
|  |  |  |  |

[rem: determine the number whose factorial yields the first sum of digits that has three digits.]

Thus, we get the simple sequence **1 2 6 6 3 9 9 9 ...,** whose sum, after five terms, also yields **9**. As unmanageable as the factorial sequence appears, this technique allows for easy checking of the results, no matter how large the factorial. Once again, these techniques cannot determine if a resulting factorial calculation is correct, however, it can immediately determine an incorrect result.

## SDQ: GENERAL PROPERTIES

Given two numbers A and B, then:

(a)  SDQ(A+B) = SDQ(A) + SDQ(B)

(b)  SDQ(A-B) = SDQ(A) - SDQ(B)

(c)  SDQ(A*B) = SDQ(A) * SDQ(B)

**Thus, for any number representable as a finite power series, the SDQ may be used to verify the correctness of that expansion. Again, the verification determines incorrectness, not necessarily correctness.**


## ADDITION QUALITY (AQ)

Another aspect of number theory is one called Theosophic Addition. In this system, number has an addition quality (AQ), given by the addition of the number together with all of its prior numbers, starting from 0 up to and including that number. Thus, AQ(4) = 10, since AQ(4) = 4+3+2+1+0= 10. It is like a factorial but instead of multiplying the sequence of numbers, they are added. [rem: look at sigma (n-1)]

Table 4-7 lists the numbers from 0 to 10 and shows how to determine the resulting quality of addition for each.

TABLE 4-7: AQ

| NUMBER | AQ |
|---|---|
| 0=0 | 0 |
| 1=0+1 | 1 |
| 2=0+1+2 | 3 |
| 3=0+1+2+3 | 6 |
| 4=0+1+2+3+4 | 10 |
| 5=0+1+2+3+4+5 | 15 |
| 6=0+1+2+3+4+5+6 | 21 |
| 7=0+1+2+3+4+5+6+7 | 28 |
| 8=0+1+2+3+4+5+6+7+8 | 36 |
| 9=0+1+2+3+4+5+6+7+8+9 | 45 |
| 10=0+1+2+3+4+5+6+7+8+9+10 | 55 |

The number 10 yields 55, ten is like an army arrayed and yields 5 over and against 5. See SY

One immediately notices from the table that once the AQ is more than one digit, we may then take the SDQ of that AQ. The first such application is for the AQ of 4 which equals 10, and then the SDQ of 10 equals 1, as seen in Table 4-8.

## TABLE 4-8 AQ AND SDQ(AQ)

| NUMBER | AQ | SDQ(AQ) |
|---|---|---|
| 0=0 | 0 | 0 |
| 1=0+1 | 1 | 1 |
| 2=0+1+2 | 3 | 3 |
| 3=0+1+2+3 | 6 | 6 |
| 4=0+1+2+3+4 | 10 | 1 |
| 5=0+1+2+3+4+5 | 15 | 6 |
| 6=0+1+2+3+4+5+6 | 21 | 3 |
| 7=0+1+2+3+4+5+6+7 | 28 | 1 |
| 8=0+1+2+3+4+5+6+7+8 | 36 | 9 |
| 9=0+1+2+3+4+5+6+7+8+9 | 45 | 9 |
| 10=0+1+2+3+4+5+6+7+8+9+10 | 55 | 1 |

## REMARKABLE RESULTS OF THE SDQ(AQ)

The remarkable results of applying the single digit quality to the quality of addition is immediately observed. The SDQ(AQ) produces the following cycle, following the first zero: 1,3,6,1,6,3,1,9,9; 1,3,6,1,6,3,1,9,9. Nine digits are in the cycle. Additionally, we see that the SDQ of the starting number, always has the following relationship to its SDQ(AQ).

**TABLE 4-9: SDQ AND SDQ(AQ)**

| NUMBER | SDQ(NUMBER) | SDQ(AQ) |
|:------:|:-----------:|:-------:|
| 0 | 0 | 0 |
| 1 | 1 | 1 |
| 2 | 2 | 3 |
| 3 | 3 | 6 |
| 4 | 4 | 1 |
| 5 | 5 | 6 |
| 6 | 6 | 3 |
| 7 | 7 | 1 |
| 8 | 8 | 9 |
| 9 | 9 | 9 |
| 10 | 1 | 1 |
| 11 | 2 | 3 |
| 12 | 3 | 6 |
| 13 | 4 | 1 |
| 14 | 5 | 6 |
| 15 | 6 | 3 |
| 16 | 7 | 1 |
| 17 | 8 | 9 |
| 18 | 9 | 9 |
| 19 | 1 | 1 |
| 20 | 2 | 3 |
| 21 | 3 | 6 |
| 22 | 4 | 1 |

We therefore find the following relationship between the SDQ of a number and the SDQ of its quality of addition, AQ, number:

SDQ          0  1  2  3  4  5  6  7  8  9
SDQ(AQ)   0  1  3  6  1  6  3  1  9  9   for all numbers.

**FINDING THE SDQ(AQ) FOR ANY NUMBER**

To find the SDQ(AQ) of any number, simply divide the number in question by 9. The remainder indicates the SDQ(AQ). Thus, consider the number 1,012. Dividing by 9 yields 112 with a remainder of 4. All we need is this remainder, to count along the 1,3,6,1,6,3,1,9,9 sequence. Thus from the following table

REMAINDER   1  2  3  4  5  6  7  8  0

SDQ(AQ)        1  3  6  1  6  3  1  9  9

we note that a remainder of 4 means an SDQ(AQ) of 1. As a further check, SDQ(1,012) = 4, and from the previous table, corresponds to 1.


**PROPERTIES OF THE SDQ(AQ)**

There are only five different SDQ(AQ) numbers 0, 1, 3, 6, and 9. In their cycle of 1 3 6 1 6 3 1 9 9 (other than the initial and only 0), the number 1 occurs three times, while there are dual appearances for the numbers 3, 6 and 9.

To the ancients 0 represents All, and 9 represents completion. Since the zero, All, and the nine, Completion, are always present, the new numbers are 1, 3 and 6. The sum of the numbers in the cycle, total 39, with an SDQ = 3. The total of the 5 different digits used here is 19, with an SDQ = 1— following the adage "Three in Aspect, One in Essence." Notice, that after the Great Female 0, there is only one other female, the 6. The 1, 3, and 9 each are male.

In each cycle therefore, there are 9 digits, or characteristics; ending with a double male completion, the 9. The cycle always originates with the great male One; then proceeds with the triune 3, and then the female 6. Again the One male continues, followed by the female 6 and then the male 3. Finally the cycle ends with the great male One and the double males 9,9.


**ADDITIONAL TESTS FOR INCORRECTNESS**

As further test for correctness, we may write each multiplier as the sum of two or more terms and test each subtotal result, together with the total. Thus, for example to test 197 x 3,045 = 599,865; we convert to SDQ and get: 8 x 3 = 24 => 6. Looking at the product, indeed 5 + 9 + 9 +  8 + 6 + 5 = 42 => 6. Suppose that the product, 599,865 was written in error as **586,995**; with digits rearranged, and last digits are the same.

Using the SDQ(AQ) test, we divide 586,995 by 9, yielding 65,221 with a remainder of 6, and this corresponds to an SDQ(AQ) value of 3. The SDQ(586,995) => 42 => 6, and this corresponds to an SDQ(AQ) of 3. NG

DIVIDING BY 9 TO FIND THE REMAINDER AND TAKING THE SDQ ARE

REM: Any numbers with the same SDQ are related, by being in the same family, like that of the periodic table. They will be in the same family with different periods. Thus as a further check we must also compare periods. The correct result, 599,865 has a period of (599,865/9) 66,651, (SDQ of period is 24 => 6) while the incorrect result, 586,995, has a period of (586,995/9) of 65,221 (SDQ of period is 16 => 7).

137 has a period of 15, and 3,045 has a period of 338.

(15 x 338 = 5,070).

REM: look at the octaves, multiples of 2!!!!!!!!!!!!!!!!

## SUM OF ODD SQUARES

Consider the following series, where we add the squares of odd numbers. Thus, $1^2 + 3^2 + 5^2 + ... + (2n+1)^2 = (n+1)(2n+1)(2n+3)/3$.

Specifically, 1 + 9 + 25 + 49 + 81 + 121 + 169 + 225 + 289 + 361 + 441 + 529 + 625 + 729 + 841 + 961 + 1,089 + 1,225 + 1,369 + 1,521 + 1,681 + 1,849 + 2,025 + 2,209 + 2,401 + 2,601 + 2,809 + 3,025 + 3,249 + 3,481 + ...

Taking the **SDQ** of the left hand side terms, we get:

**1 + 9 + 7 + 4 + 9 + 4 + 7 + 9 + 1 + 1 + 9 + 7 + 4 + 9 + 4 + 7 + 9 + 1 + 1 + 9 + 7 + 4 + 9 + 4 + 7 + 9 + 1 + 1 + 9 + 7 +** ... as a nine term repeating series and they form a symmetry, since we have the same numbers in reverse order.

If now we add the terms in this **SDQ** series, forming another **SDQ** series of SUMS, we get the following terms:

**1 1 8 3 3 7 5 5 6 7 7 5 9 9 4 2 2 3 4 4 2 6 6 1 8 8 9**, as a **27** term repeating series.

**Notice that the double digits start at 1 and are separated by a deceasing set of integers starting at 8, and ending at 9, (underlined numbers). The double digits increase by 2, e.g. 1 1 3 3 5 5 7 7 9 9 2 2 4 4 6 6 8 8. Thus, each of the nine digits is used three times (3x9=27 terms).**

In Table 4-10 we examine the **SDQ** of the right hand side, we get the same results as the left hand side.

**TABLE 4-10**

| n | 0 | 1 | 2 | 3 | 4 | 5 | 6 | 7 | 8 |
|---|---|---|---|---|---|---|---|---|---|
| (n+1)(2n+1)(2n+3)/3 | 1 | 10 | 35 | 84 | 165 | 286 | 455 | 680 | 969 |
| SDQ[(n+1)(2n+1)(2n+3)/3] | 1 | 1 | 8 | 3 | 3 | 7 | 5 | 5 | 6 |

We see that the **SDQ** yields a pattern that is otherwise not present in the original series, and that pattern is the same in the right hand side, or single term series sum representation.

## SUM OF ODD CUBES

Consider now the series of the addition the cubes of odd integers. Thus, $1^3 + 3^3 + 5^3 + ... + (2n+1)^3 = (n+1)^2 (2n^2+4n+1)$. Taking the **SDQ** of the left hand side, yields: **1 + 9 + 8** + 1 + 9 + 8, a three term repeating series. If we now evaluate the **SDQ** of the sum, we get the following terms: **1 1 9**, 1 1 9, ...
The right hand side yields the same results, **1 1 9**.

**TABLE 4-11**

| n | 0 | 1 | 2 | 3 | 4 | 5 |
|---|---|---|---|---|---|---|
| $(n+1)^2$ x $(2n^2+4n+1)$ | 1 | 28 | 153 | 496 | 1225 | 2556 |
| SDQ[$(n+1)^2$ x$(2n^2+4n+1)$] | 1 | 1 | 9 | 1 | 1 | 9 |

## SUMMARY

We see therefore, that number has played many roles, as qualities, as quantities, and as potencies. One such quality is that obtained by successive additions of the digits. When reduced to a single digit, there are nine possibilities, in addition to the zero. Another quality is that or sex, even numbers being female, and odd numbers, male. At each stage of digit reduction, number has sex. The smaller the number, the greater the potency of that number. Thus, the most potent number is 0, Female. Notice that the 0 has three regions of form; the finite inside, the semi infinite outside, and the boundary. Zero, is unique, and later we will discuss some of its properties.

EXAMPLE
    Consider the number 219:

        As a <u>quantity</u>, it represents 219 objects or things.
        As a <u>potency</u>, it equals 219, or 2+1+9 = 12, or 1+2 =3. It therefore has
            three potencies; 219, 12 and 3.
        It has the <u>quality</u> of 3.
        It has the sexual qualities of; 219 primary Male, 12 secondary
        female, and 3 tertiary male. We therefore write its sexual
    qualities as Male female male, or   simply Mfm.
        It has the <u>quantity</u> of 0+1+2+...+219 = 24,090.
        Which has the <u>potencies</u> of 24,090, 15, and 6.
        Which has the sexual <u>qualities</u> of Fmf.
        Which has the <u>quality</u> of 6.
        Which has the <u>quantity</u> of 219!
        Which has the <u>potencies</u> of all of the digit sums of  219!
        Which has the <u>quality</u> of 9.


## NEGATIVE EXPONENTIALS

FAR]

$SDQ(2^{-n}) = 1, 5, 7, 8, 4, 2;$

and since $SDQ(2^n) = 1, 2, 4, 8, 7, 5;$

then, $SDQ(2^{-n}) \times SDQ(2^n) = 1$, as it should.

Also,  $SDQ(2^n) + SDQ(2^{-n}) = 2, 7; 2, 7; 2, 7;$ which equals $SDQ(2^n + 2^{-n})$.

$SDQ(4^{-n}) = 1, 7, 4; 1, 7, 4;$ [check]

$SDQ(4^n) = 1, 4, 7; 1, 4, 7;$

thus, $SDQ(4^n) + SDQ(4^{-n}) = 2; 2; 2; 2;$


$SDQ(5^n) = 1, 5, 7, 8, 4, 2; = SDQ(2^{-n})$

$SDQ(5^{-n}) = 1, 2, 4, 8, 7, 5; = SDQ(2^n)$

Thus, as with the $2^n$, $SDQ(5^n) + SDQ(5^{-n}) = 2, 7; 2, 7;$

$SDQ(8^n) = SDQ(8^{-n}) = 1, 8;$

and the sum of the SDQ's are: 2, 7; 2, 7; as with the fives.

**TABLE 4-12**

| n | $2^n$ | SDQ($2^n$) | $2^{-n}$ | $2^{-n}$ | SDQ($2^{-n}$) |
|---|---|---|---|---|---|
| 0 | 1 | 1 | 1/1 | 1.0 | 1 |
| 1 | 2 | 2 | 1/2 | 0.5 | 5 |
| 2 | 4 | 4 | 1/4 | 0.25 | 7 |
| 3 | 8 | 8 | 1/8 | 0.125 | 8 |
| 4 | 16 | 7 | 1/16 | 0.0625 | 4 |
| 5 | 32 | 5 | 1/32 | 0.03125 | 2 |
| | | | | | |
| 6 | 64 | 1 | 1/64 | 0.015625 | 1 |
| 7 | 128 | 2 | 1/128 | 0.0078125 | 5 |

**TABLE 4-13**

| n | $4^n$ | SDQ($2^n$) | $(4^{(-n)})$ | $4^{(-n)}$ | SDQ($4^{(-n)}$) |
|---|---|---|---|---|---|
| 0 | 1 | 1 | 1/1 | 1 | 1 |
| 1 | 4 | 4 | 1/4 | 0.25 | 7 |
| 2 | 16 | 7 | 1/16 | 0..0625 | 4 |
| 3 | 64 | 1 | 1/64 | 0.015625 | 1 |
| 4 | 256 | 4 | 1/256 | 0.00390625 | 7 |
| 5 | 1024 | 7 | 1/1024 | 0.0009765625 | 4 |
| | | | | | |
| 6 | 4096 | 1 | 1/4096 | | 1 |
| 7 | 16384 | 4 | 1/16384 | | 7 |

We note that since $(2^n) \times (2^{-n}) = 1$, then SDQ $(2^n) \times$ SDQ $(2^{-n}) = 1$.

From the minimum SDQ multiplication table 3, we see that the numbers that yield an SDQ of 1 are:

# Chapter 5

# Fibonacci, The Golden Proportion and SDQ

**Fibonacci Growth**

It has been shown that patterns of growth can be exhibited mathematically by the Fibonacci series, a recursive sequence where each term is the sum of its two preceding terms. The Fibonacci series $F_{n+2} = F_{n+1} + F_n$ where $F_1 = 1$, $F_2 = 1$ and $n = 1, 2, 3 \ldots$

We shall refer to this well known series as "Fibonacci" or "Bi-Fi," short for "Bi-Fibonacci" to distinguish it from a "Tri-Fi" series, characterized by each term being the sum of its three preceding terms. Tri-Fi and larger preceding term sum Fibonacci type series will be examined in this work.

The typical Bi-Fi series is: 1, 1, 2, 3, 5, 8, 13, 21, 34, 55, 89, 144, 233, 377, 610, 987, 1,597, 2,584, 4,181, 6,765, 10,946, 17,711, 28,657, 46,368, 75,025, 121,393, 196,418, 317,811, 514,229, 832,040… We see that $1 + 1 = 2$, $1 + 2 = 3$, $2 + 3 = 5$, $3 + 5 = 8$, $8 + 5 = 13$, and so forth.

A piano keyboard had 5 black keys, 8 white keys, and 13 black and white keys which together make up the octave, following the Fibonacci sequence rule. As we move out on the sequence, the ratio of their terms approaches the Golden Proportion, 1.618, while their inverse ratio approaches 0.618. Thus, even using the sixth and seventh terms, 8 and 13, we get that $13/8 = 1.625$ and $8/13 = 0.615$. 13 divided by 8 differs from Phi, $\phi$ (1.618) by less than one half of one percent.

The number, 1.618 is known as Phi, $\phi$ and has the significant property that $1 + \phi = \phi^2$. It is possible to start the Fibonacci series either with 0 and 1, or with 1 and 1, or with 1 and 2; each yielding the same resulting series. Starting with the dual 1 and 1, we get the following: **1, 1, 2, 3, 5, 8, 13, 21, 34, 55, 89, 144, 233, 377, 610, 987, 1,597, 2,584, 4,181, 6,765, 10,946, 17,711, 28,657, 46,368;** 75,025, 121,393, 196,418, 317,811, 514,229, 832,040…

We have placed a semicolon after the twenty-fourth term, 46,368 because we shall show that when each term is reduced to a single digit, known as Single Digit Quality SDQ or mod 9, then 43,368 is the last term of a twenty-four term repeating cycle. We mark the start of the next cycle by italics. The twenty-fifth term divided by the twenty-fourth term is $75,025/46,368 = 1.610834\ldots$ the golden proportion, $\phi$ phi. The twenty-fourth term divided by the twenty-fifth term is $46,368/75,025 = 0.618034$, $\phi^{-1}$. Notice that $1 + \phi^{-1} = \phi$, and that $1 + \phi = \phi^2$.

## SEX

The sex of the individual terms, (even numbers are female, and odd numbers are male), are: male, male, female; male, male, female, . This is certainly expected due to the additive properties of odd and even numbers; namely: odd + odd = even; even + even = even, and odd + even = odd. These correspond sexually to: male + male = female; female + female = female; and male + female = male. If we start the series with 0, 1, 1, 2, 3, 5, 8, 13, 21, 34, ..., then the sexes are female, male, male; female, male, male; ... While, starting with 1, 2, 3, 5, 8, 13, 21, 34, . we get for the sexes: male, female, male; male, female, male; male, female, male; . We shall show that starting with 0 is special, and it holds a special place in the SDQ pattern.

## FIBONACCI AND SDQ

If we take the SDQ of each of the Fibonacci series terms, the results are: 1, 1, 2, 3, 5, 8, 4, 3, 7, 1, 8, 9, 8, 8, 7, 6, 4, 1, 5, 6, 2, 8, 1, 9; 1, 1, 2 , 3, 5, 8, ...   and we note that this SDQ Fibonacci series, unlike the regular Fibonacci series, is cyclic, with a cycle of 24 terms. These results are shown in Table 1.

**TABLE 1. FIBONACCI SERIES AND SDQ OF INDIVIDUAL TERMS**

| FIBONACCI SERIES | | | | | | | | | | | | | | | |
|---|---|---|---|---|---|---|---|---|---|---|---|---|---|---|---|
| Num | 1 | 1 | 2 | 3 | 5 | 8 | 13 | 21 | 34 | 55 | 89 | 144 | 233 | 377 | 610 | 987 |
| SDQ | 1 | 1 | 2 | 3 | 5 | 8 | 4 | 3 | 7 | 1 | 8 | 9 | 8 | 8 | 7 | 6 |

| FIBONACCI SERIES - CONTINUED | | | | | | | |
|---|---|---|---|---|---|---|---|
| Num | 1,597 | 2,584 | 4,181 | 6,765 | 10,946 | 17,711 | 28,657 | 46,368 |
| SDQ | 4 | 1 | 5 | 6 | 2 | 8 | 1 | 9 |

| FIBONACCI SERIES - CONTINUED | | | | | |
|---|---|---|---|---|---|
| Num | 75,025 | 121,393 | 196,418 | 317,811 | 514,229 | 832,040 |
| SDQ | 1 | 1 | 2 | 3 | 5 | 8 |

## PROPERTIES OF SDQ FIBONACCI

In each cycle then, there are: five 1's; two 2's; two 3's; two 4's; two 5's; two 6's; two 7's; five 8's and two 9's; five 1's and 8's, and two each of every other digit.

The sum of the SDQ cycle is 117 => **9**. The sum of the first 24 terms of the regular Fibonacci series equals 121,392 => **9**.

The Fibonacci series upon which the Golden Proportion rests, when reduced through SDQ, is a repetitive series with period, P = 24. Note the period of 24 => 6.

N.B. Starting with 1 and 1, the tenth term of the Fibonacci series is a 55. If we start with 0 and 1, then it is the 11th term. If we start the series with a 1 and 2, then it is the 9th term. Ten not eleven, ten and not nine, for 5 over against 5, could mean that we start with the dual 1 and then 55 becomes the tenth term. (See SY).

## FAST CALCULATION OF FIBONACCI SDQ

In order to calculate the SDQ of each term, we need not carry out the full addition, but can just work with SDQ of each term and when we add, always reduce to the SDQ. Thus, we can immediately write the terms as 1, 1, 2, 3, 5, 8, 4 (since 8 + 5 = 13 => 4) and so forth. This is extremely convenient, and allows a lightening calculation of each term SDQ for any series. We will find that every Fibonacci type series, starting with any two terms, which are then added to produce the next term, and so forth, always has an SDQ cycle of 24 terms (we will not start with 0, as there is no other SDQ that produces 0).

## Simplification

The SDQ of the Fibonacci Series is 1, 1, 2, 3, 5, 8, 4, 3, 7, 1, 8, 9, 8, 7, 6, 4, 1, 5, 6, 2, 8, 1, 9;

If we examine the first half of the 24 term series, 1, 1, 2, 3, 5, 8, 4, 3, 7, 1, 8, 9 and compare it with the second half of this series, 8, 8, 7, 6, 4, 1, 5, 6, 2, 8, 1, 9, writing the second half directly below the first.

1, 1, 2, 3, 5, 8, 4, 3, 7, 1, 8, 9
8, 8, 7, 6, 4, 1, 5, 6, 2, 8, 1, 9;

Every set of upper and lower terms add to 9. (The last set of 9 and 9 add to 18 which has and SDQ of 9.) We see that the full series only requires 12 terms and the next 12 are immediately derivable from the first set.

The SDQ of the Fibonacci Sequence is 1 1 2 3 5 8 4 3 7 1 8 9 8 8 7 6 4 1 5 6 2 8 1 9;
In reversed order, it is  9 1 8 2 6 5 1 4 6 7 8 8 9 8 1 7 3 4 8 5 3 2 1 1; whose sum SDQ becomes: 1 9 2 8 4 5 9 6 4 3 2 2 1 2 9 3 7 6 2 5 7 8 9 9;

Adding the reverse order of the SDQ of the Fibonacci Sequence to SDQ to its order reversed sum and listing them one above the other:

9 1 8 2 6 5 1 4 6 7 8 8 9 9 8 1 7 3 4 8 5 3 2 1 1
1 1 2 3 5 8 4 3 7 1 8 9 8 8 7 6 4 1 5 6 2 8 1 9

9 1 8 2 6 5 1 4 6 7 8 8 9 9 8 1 7 3 4 8 5 3 2 1 1
1 9 2 8 4 5 9 6 4 3 2 2 1 2 9 3 7 6 2 5 7 8 9 9

## A Fibonacci named Moebius

A Moebius strip is a one sided strip that is given a twist and then connected end to end. In this fashion two dimensional becomes one dimensional since a walk along the connected strip will take cover both sides.

We can write the SDQ Fibo numbers on a Moebius strip and it then becomes a repeating pattern. To make the strip we need only use the first 12 terms and then the next 12 are written upside down and opposite the first set. Then this strip is twisted and connected. Every number on one side when added to its mate on the other side adds to 9. It is the true meaning of the infinity sign which is continuous and unending.

[FAR}

[Could this SDQ Strip numbering be applicable to the physics of String Theory?]

David Benbennick: March 14, 2005

This could also be representative of a two year cycle of 12 months each.

**Fibonacci Decay**

Fibonacci growth patterns are given by the Fibonacci series $F_{n+2} = F_{n+1} + F_n$ where $F_1 = 1$, $F_2 = 1$ and $n = 1, 2, 3 \ldots$ The SDQ of the Fibonacci growth pattern is 1, 1, 2, 3, 5, 8, 4, 3, 7, 1, 8, 9, 8, 8, 7, 6, 4, 1, 5, 6, 2, 8, 1, 9;

Fibonacci decay patterns are given by $SDQ(F_{n+2}) = SDQ(F_{n+1}) - SDQ(F_n)$, where $F_1 = 9$, $F_2 = 1$ and $n = 1, 2, 3 \ldots$ This is a recursive sequence with each term is the difference between the prior and next prior terms.

This is shown by writing the SDQ growth pattern in reverse, namely:
9, 1, 8, 2, 6, 5, 1, 4, 6, 7, 8, 8, 9, 8, 1, 7, 3, 4, 8, 5, 3, 2, 1, 1; again a 24 term repeating sequence.
For this SDQ (mod 9) sequence, we use the relationships; $n - n = 9$, and $n - 9 = n$, for any n.

The corresponding terms in the second half of the sequence when placed below the first half, add to 9. Thus,
9, 1,  8, 2, 6, 5, 1, 4, 6, 7, 8, 8
9, 8, 1, 7, 3, 4, 8, 5, 3, 2, 1, 1 and therefore we need only calculate the first half of the sequence to immediately determine the following corresponding twelve terms.

**Fibonacci Continuing Sum Growth SDQ**

The continuing sum of all preceding terms is given by: 1, 2, 4, 7, 12, 20, 33, 54, 88, 143, 232, 376, 609, ... . A quick method of summing is to begin with 1, and then starting with the third term in the regular series decrease subsequent terms by 1, namely: 1, 2, 4, 7, 12, 20, 33, 54, 88, 143, 232, 376, 609, 986, 1596, 2583, 4180, 6764, 10945, 17710, 28656, 46367, 75024, 121392, 196417, ...
The ratios of terms yield the same golden proportions $\phi$ and $\phi^{-1}$.

The SDQ of this continuing sum is: 1, 2, 4, 7, 3, 2, 6, 9, 7, 8, 7, 7, 6, 5, 3, 9, 4, 5, 1, 7, 9, 8, 9, 9; 1, 2 again a 24 term repeating series. It has Period sum SDQ = 138 => 3. Again if we separate the first twelve terms and place the next twelve directly below,
1, 2, 4,  7, 3, 2, 6, 9,  7, 8, 7, 7,
6, 5, 3, 9, 4, 5, 1, 7, 9, 8, 9, 9 and note that we do not have to calculate the second set of terms. Whereas in the regular series corresponding terms add to 9, here they add to 7.

**Fibonacci Continuing Sum Decay SDQ**

The continuing sum decay SDQ is the reverse of the sum growth SDQ namely: 9, 9, 8, 9, 7, 1, 5, 4, 9, 3, 5, 6, 7, 7, 8, 7, 9, 6, 2, 3, 7, 4, 2, 1.

When added to the SDQ of the Fibonacci growth pattern: 1, 1, 2, 3, 5, 8, 4, 3, 7, 1, 8, 9, 8, 8, 7, 6, 4, 1, 5, 6, 2, 8, 1, 9 shown directly below one another, becomes:

9, 9, 8, 9, 7, 1, 5, 4, 9, 3, 5, 6, 7, 7, 8, 7, 9, 6, 2, 3, 7, 4, 2, 1

1, 1, 2, 3, 5, 8, 4, 3, 7, 1, 8, 9, 8, 8, 7, 6, 4, 1, 5, 6, 2, 8, 1, 9

given by: 1, 2, 4, 7, 12, 20, 33, 54, 88, 143, 232, 376, 609, ... . Just start with the third term in the regular series and then decrease each term by 1. 1, 2, 4, 7, 12, 20, 33, 54, 88, 143, 232, 376, 609, 986, 1596, 2583, 4180, 6764, 10945, 17710, 28656, 46367, 75024, 121392, 196417, ... . The SDQ of the repeating sum is: 1, 2, 4, 7, 3, 2, 6, 9, 7, 8, 7, 7, 6, 5, 3, 9, 4, 5, 1, 7, 9, 8, 9, 9; 1, 2 again a 24 repeating series. Thus, the sum is always 1 less than the term after the one in question. It has the sum of the SDQ = 138 => 3.

Thus we see that taking the reversed order of the FS single digits, and then doing an SDQ on them, then adding to the reversed order FS single digits, produces 10 for each sum of corresponding digits.

; In reversed order, it is ; whose SDQ becomes: 1 9 2 8 4 5 9 6 4 3 2 2 1 2 9 3 7 6 2 5 7 8 9 9;

Adding the SDQ of the reversed order to the reversed order  9 1 8 2 6 5 1 4 6 7 8 8 9 8 1 7 3 4 8 5 3 2 1 1.

## FIBONACCI -LIKE SEQUENCE

Consider a Fibonacci like sequence that starts with the numbers 1 and 2; thus 1, 2, 3, 5, 8, 13, 21, 34, 55, 89, ...  If we add the first twelve terms of this Fibonacci like sequence, terms (1+2+3+ ... +89), the total is 231. This number, 231, is the same as the number of ways that 22 Hebrew letters can be arranged taking two at a time. **[rem: simplify] e.g. the Hebrew letters. 22x21/2 = 231.**

## THE LUCAS SEQUENCE

This sequence is closely related to the Fibonacci sequence in that again the next term in the sequence is the sum of the two prior terms, $L_n = L_{n-1} + L_{n-2}$ where $L_1 = 1$ and $L_2 = 3$. Here we start with the numbers 1 and 3.

1, 3, 4, 7, 11, 18, 29, 47, 76, 123, 199, 322, 521, 843, 1364, 2207, 3571, 5778, 9349, 15127, 24476, 39603, 64079, 103682; 167761, 271443, 439204, ...

If we take the ratio of a term to its preceding term we get 439204/271443 =1.618+ or φ.

## SDQ
1, 3, 4, 7, 2, 9, 2, 2, 4, 6, 1, 7, 8, 6, 5, 2, 7, 9, 7, 7, 5, 3, 8, 2; 1, 3,...

Once again, we get a 24 term repeating SDQ series. Here there are: Two 1's; five 2's; two 3's; two 4's; two 5's; two 6's; five 7's; two 8's; and two 9's. Once again: five 2's, five 7's, and two each of the other digits. We also notice that when the first term is added to the 13th term, and this sequence continues, as with Fibonacci, each set add to 9.

## ANOTHER FIBONACCI-LIKE SERIES: ALL EVEN (FEMALE)
2, 2, 4, 6, 10, 16, 26, 42, 68, 110, 178, 288, 466, 754, 1220, 1974, 3194, 5168, 8362, 13530, 21892, 35422, 57314, 92736; 150050, 242786,

If we take the ratio of a term to its preceding term we get 242786/150050 = 1.618+ or .
2, 2, 4, 6, 1, 7, 8, 6, 5, 2, 7, 9, 7, 7, 5, 3, 8, 2, 1, 3, 4, 7, 2, 9; 2, 2,

N.B. (note well): there are two ways of arriving at the SDQ of a sequence; either add the terms and then reduce to the SDQ, or determine the SDQ of each term, then add these terms, and perform the final resulting SDQ. Again a 24 term repeating series. five 2's and 7's, and two each of the other digits. The SDQ of the first 24 terms add to 117 => 9.

## STILL ANOTHER FIBONACCI-LIKE SERIES
9, 9, 18, 27, 45, 72, 117, 189, 306, 495, 801, 1296,
and we see that every SDQ here is a 9.

If we take the ratio of term to its preceding term we get 1296/801 = 1.61797 approaching φ.

None of the terms of the series is a prime number, as there cannot be a prime with an SDQ of 3, 6, or 9, and in this case, all of the SDQ are 9.

## GENERAL FIBONACCI-LIKE SERIES
The general Fibonacci-like series is of the form: a, b, a + b, a + 2b, 2a + 3b, 3a + 5b, 5a + 8b, 8a + 13b, 13a + 21b, 21a + 34b, 34a + 55b, 55a + 89b, 89a + 144b, 144a + 233b, 233a + 377b, 377a + 610b, 610a + 987b, 987a + 1597b, 1597a + 2584b,

2584a + 4181b, 4181a + 6765b, 6765a + 10946b, 10946a + 17711b, 17711a + 28657b; 28657a + 46368b, 46368a + 75025b, 75025a + 121393b, 121393a + 196418b,

## SDQ OF GENERAL FIBONACCI-LIKE SERIES

Since the SDQ of a product equals the product of the SDQ, we take the SDQ of each of the coefficients of a and b. This equals: a, b, a + b, a + 2b, 2a + 3b, 3a + 5b, 5a + 8b, 8a + 4b, 4a + 3b, 3a + 7b, 7a + 1b, 1a + 8b, 8a + 9b, 9a + 8b, 8a + 8b, 8a + 7b, 7a + 6b, 6a + 4b, 4a + 1b, 1a + 5b, 5a + 6b, 6a + 2b, 2a + 8b, 8a + 1b; 1a + 9b, 9a + 1b, 1a + 1b, 1a + 2b, ... . Since SDQ(1a + 9b) => SDQ(a), and SDQ(9a + 1b) => SDQ(b), we see that every Fibonacci-like series has a period of 24 terms. Note that SDQ (24) => 6.

Because we are using SDQ, we find that we can calculate any set of starting numbers in the Fibonacci Series by using 9 tables. We will illustrate these tables and note the interesting features that appear. Each is a 24 term repeating series and each table shows the first 26 terms. We start with 1 as the first number, and proceed from 1 to 9. We stop at 9, since SDQ of 10 is the same as 1 which brings us back to the first table.

**1,1; 1,2; 1,3; etc.**

| TWENTY FOUR TERM REPEATING SERIES | | | | | | | | | | | | | | | | | | | | | | | | | |
|---|---|---|---|---|---|---|---|---|---|---|---|---|---|---|---|---|---|---|---|---|---|---|---|---|---|
| 1 | 1 | 2 | 3 | 5 | 8 | 4 | 3 | 7 | 1 | 8 | 9 | 8 | 8 | 7 | 6 | 4 | 1 | 5 | 6 | 2 | 8 | 1 | 9 | 1 | 1 |
| 1 | 2 | 3 | 5 | 8 | 4 | 3 | 7 | 1 | 8 | 9 | 8 | 8 | 7 | 6 | 4 | 1 | 5 | 6 | 2 | 8 | 1 | 9 | 1 | 1 | 2 |
| 1 | 3 | 4 | 7 | 2 | 9 | 2 | 2 | 4 | 6 | 1 | 7 | 8 | 6 | 5 | 2 | 7 | 9 | 7 | 7 | 5 | 3 | 8 | 2 | 1 | 3 |
| 1 | 4 | 5 | 9 | 5 | 5 | 1 | 6 | 7 | 4 | 2 | 6 | 8 | 5 | 4 | 9 | 4 | 4 | 8 | 3 | 2 | 5 | 7 | 3 | 1 | 4 |
| 1 | 5 | 6 | 2 | 8 | 1 | 9 | 1 | 1 | 2 | 3 | 5 | 8 | 4 | 3 | 7 | 1 | 8 | 9 | 8 | 8 | 7 | 6 | 4 | 1 | 5 |
| 1 | 6 | 7 | 4 | 2 | 6 | 8 | 5 | 4 | 9 | 4 | 4 | 8 | 3 | 2 | 5 | 7 | 3 | 1 | 4 | 5 | 9 | 5 | 5 | 1 | 6 |
| 1 | 7 | 8 | 6 | 5 | 2 | 7 | 9 | 7 | 7 | 5 | 3 | 8 | 2 | 1 | 3 | 4 | 7 | 2 | 9 | 2 | 2 | 4 | 6 | 1 | 7 |
| 1 | 8 | 9 | 8 | 8 | 7 | 6 | 4 | 1 | 5 | 6 | 2 | 8 | 1 | 9 | 1 | 1 | 2 | 3 | 5 | 8 | 4 | 3 | 7 | 1 | 8 |
| 1 | 9 | 1 | 1 | 2 | 3 | 5 | 8 | 4 | 3 | 7 | 1 | 8 | 9 | 8 | 8 | 7 | 6 | 4 | 1 | 5 | 6 | 2 | 8 | 1 | 9 |

**2,1; 2,2; 2,3; etc.**

| TWENTY FOUR TERM REPEATING SERIES |||||||||||||||||||||||||| |
|---|---|---|---|---|---|---|---|---|---|---|---|---|---|---|---|---|---|---|---|---|---|---|---|---|---|
| 2 | 1 | 3 | 4 | 7 | 2 | 9 | 2 | 2 | 4 | 6 | 1 | 7 | 8 | 6 | 5 | 2 | 7 | 9 | 7 | 7 | 5 | 3 | 8 | 2 | 1 |
| 2 | 2 | 4 | 6 | 1 | 7 | 8 | 6 | 5 | 2 | 7 | 9 | 7 | 7 | 5 | 3 | 8 | 2 | 1 | 3 | 4 | 7 | 2 | 9 | 2 | 2 |
| 2 | 3 | 5 | 8 | 4 | 3 | 7 | 1 | 8 | 9 | 8 | 8 | 7 | 6 | 4 | 1 | 5 | 6 | 2 | 8 | 1 | 9 | 1 | 1 | 2 | 3 |
| 2 | 4 | 6 | 1 | 7 | 8 | 6 | 5 | 2 | 7 | 9 | 7 | 7 | 5 | 3 | 8 | 2 | 1 | 3 | 4 | 7 | 2 | 9 | 2 | 2 | 4 |
| 2 | 5 | 7 | 3 | 1 | 4 | 5 | 9 | 5 | 5 | 1 | 6 | 7 | 4 | 2 | 6 | 8 | 5 | 4 | 9 | 4 | 4 | 8 | 3 | 2 | 5 |
| 2 | 6 | 8 | 5 | 4 | 9 | 4 | 4 | 8 | 3 | 2 | 5 | 7 | 3 | 1 | 4 | 5 | 9 | 5 | 5 | 1 | 6 | 7 | 4 | 2 | 6 |
| 2 | 7 | 9 | 7 | 7 | 5 | 3 | 8 | 2 | 1 | 3 | 4 | 7 | 2 | 9 | 2 | 2 | 4 | 6 | 1 | 7 | 8 | 6 | 5 | 2 | 7 |
| 2 | 8 | 1 | 9 | 1 | 1 | 2 | 3 | 5 | 8 | 4 | 3 | 7 | 1 | 8 | 9 | 8 | 8 | 7 | 6 | 4 | 1 | 5 | 6 | 2 | 8 |
| 2 | 9 | 2 | 2 | 4 | 6 | 1 | 7 | 8 | 6 | 5 | 2 | 7 | 9 | 7 | 7 | 5 | 3 | 8 | 2 | 1 | 3 | 4 | 7 | 2 | 9 |

**3,1; 3,2; 3,3; etc.**

| TWENTY FOUR TERM REPEATING SERIES |||||||||||||||||||||||||| |
|---|---|---|---|---|---|---|---|---|---|---|---|---|---|---|---|---|---|---|---|---|---|---|---|---|---|
| 3 | 1 | 4 | 5 | 9 | 5 | 5 | 1 | 6 | 7 | 4 | 2 | 6 | 8 | 5 | 4 | 9 | 4 | 4 | 8 | 3 | 2 | 5 | 7 | 3 | 1 |
| 3 | 2 | 5 | 7 | 3 | 1 | 4 | 5 | 9 | 5 | 5 | 1 | 6 | 7 | 4 | 2 | 6 | 8 | 5 | 4 | 9 | 4 | 4 | 8 | 3 | 2 |
| 3 | 3 | 6 | 9 | 6 | 6 | 3 | 9 | 3 | 3 | 6 | 9 | 6 | 6 | 3 | 9 | 3 | 3 | 6 | 9 | 6 | 6 | 3 | 9 | 3 | 3 |
| 3 | 4 | 7 | 2 | 9 | 2 | 2 | 4 | 6 | 1 | 7 | 8 | 6 | 5 | 2 | 7 | 9 | 7 | 7 | 5 | 3 | 8 | 2 | 1 | 3 | 4 |
| 3 | 5 | 8 | 4 | 3 | 7 | 1 | 8 | 9 | 8 | 8 | 7 | 6 | 4 | 1 | 5 | 6 | 2 | 8 | 1 | 9 | 1 | 1 | 2 | 3 | 5 |
| 3 | 6 | 9 | 6 | 6 | 3 | 9 | 3 | 3 | 6 | 9 | 6 | 6 | 3 | 9 | 3 | 3 | 6 | 9 | 6 | 6 | 3 | 9 | 3 | 3 | 6 |
| 3 | 7 | 1 | 8 | 9 | 8 | 8 | 7 | 6 | 4 | 1 | 5 | 6 | 2 | 8 | 1 | 9 | 1 | 1 | 2 | 3 | 5 | 8 | 4 | 3 | 7 |
| 3 | 8 | 2 | 1 | 3 | 4 | 7 | 2 | 9 | 2 | 2 | 4 | 6 | 1 | 7 | 8 | 6 | 5 | 2 | 7 | 9 | 7 | 7 | 5 | 3 | 8 |
| 3 | 9 | 3 | 3 | 6 | 9 | 6 | 6 | 3 | 9 | 3 | 3 | 6 | 9 | 6 | 6 | 3 | 9 | 3 | 3 | 6 | 9 | 6 | 6 | 3 | 9 |

4,1; 4,2; 4,3; etc.

| TWENTY FOUR TERM REPEATING SERIES | | | | | | | | | | | | | | | | | | | | | | | | | |
|---|---|---|---|---|---|---|---|---|---|---|---|---|---|---|---|---|---|---|---|---|---|---|---|---|---|
| 4 | 1 | 5 | 6 | 2 | 8 | 1 | 9 | 1 | 1 | 2 | 3 | 5 | 8 | 4 | 3 | 7 | 1 | 8 | 9 | 8 | 8 | 7 | 6 | 4 | 1 |
| 4 | 2 | 6 | 8 | 5 | 4 | 9 | 4 | 4 | 8 | 3 | 2 | 5 | 7 | 3 | 1 | 4 | 5 | 9 | 5 | 5 | 1 | 6 | 7 | 4 | 2 |
| 4 | 3 | 7 | 1 | 8 | 9 | 8 | 8 | 7 | 6 | 4 | 1 | 5 | 6 | 2 | 8 | 1 | 9 | 1 | 1 | 2 | 3 | 5 | 8 | 4 | 3 |
| 4 | 4 | 8 | 3 | 2 | 5 | 7 | 3 | 1 | 4 | 5 | 9 | 5 | 5 | 1 | 6 | 7 | 4 | 2 | 6 | 8 | 5 | 4 | 9 | 4 | 4 |
| 4 | 5 | 9 | 5 | 5 | 1 | 6 | 7 | 4 | 2 | 6 | 8 | 5 | 4 | 9 | 4 | 4 | 8 | 3 | 2 | 5 | 7 | 3 | 1 | 4 | 5 |
| 4 | 6 | 1 | 7 | 8 | 6 | 5 | 2 | 7 | 9 | 7 | 7 | 5 | 3 | 8 | 2 | 1 | 3 | 4 | 7 | 2 | 9 | 2 | 2 | 4 | 6 |
| 4 | 7 | 2 | 9 | 2 | 2 | 4 | 6 | 1 | 7 | 8 | 6 | 5 | 2 | 7 | 9 | 7 | 7 | 5 | 3 | 8 | 2 | 1 | 3 | 4 | 7 |
| 4 | 8 | 3 | 2 | 5 | 7 | 3 | 1 | 4 | 5 | 9 | 5 | 5 | 1 | 6 | 7 | 4 | 2 | 6 | 8 | 5 | 4 | 9 | 4 | 4 | 8 |
| 4 | 9 | 4 | 4 | 8 | 3 | 2 | 5 | 7 | 3 | 1 | 4 | 5 | 9 | 5 | 5 | 1 | 6 | 7 | 4 | 2 | 6 | 8 | 5 | 4 | 9 |

5,1; 5,2; 5,3; etc.

| TWENTY FOUR TERM REPEATING SERIES | | | | | | | | | | | | | | | | | | | | | | | | | |
|---|---|---|---|---|---|---|---|---|---|---|---|---|---|---|---|---|---|---|---|---|---|---|---|---|---|
| 5 | 1 | 6 | 7 | 4 | 2 | 6 | 8 | 5 | 4 | 9 | 4 | 4 | 8 | 3 | 2 | 5 | 7 | 3 | 1 | 4 | 5 | 9 | 5 | 5 | 1 |
| 5 | 2 | 7 | 9 | 7 | 7 | 5 | 3 | 8 | 2 | 1 | 3 | 4 | 7 | 2 | 9 | 2 | 2 | 4 | 6 | 1 | 7 | 8 | 6 | 5 | 2 |
| 5 | 3 | 8 | 2 | 1 | 3 | 4 | 7 | 2 | 9 | 2 | 2 | 4 | 6 | 1 | 7 | 8 | 6 | 5 | 2 | 7 | 9 | 7 | 7 | 5 | 3 |
| 5 | 4 | 9 | 4 | 4 | 8 | 3 | 2 | 5 | 7 | 3 | 1 | 4 | 5 | 9 | 5 | 5 | 1 | 6 | 7 | 4 | 2 | 6 | 8 | 5 | 4 |
| 5 | 5 | 1 | 6 | 7 | 4 | 2 | 6 | 8 | 5 | 4 | 9 | 4 | 4 | 8 | 3 | 2 | 5 | 7 | 3 | 1 | 4 | 5 | 9 | 5 | 5 |
| 5 | 6 | 2 | 8 | 1 | 9 | 1 | 1 | 2 | 3 | 5 | 8 | 4 | 3 | 7 | 1 | 8 | 9 | 8 | 8 | 7 | 6 | 4 | 1 | 5 | 6 |
| 5 | 7 | 3 | 1 | 4 | 5 | 9 | 5 | 5 | 1 | 6 | 7 | 4 | 2 | 6 | 8 | 5 | 4 | 9 | 4 | 4 | 8 | 3 | 2 | 5 | 7 |
| 5 | 8 | 4 | 3 | 7 | 1 | 8 | 9 | 8 | 8 | 7 | 6 | 4 | 1 | 5 | 6 | 2 | 8 | 1 | 9 | 1 | 1 | 2 | 3 | 5 | 8 |
| 5 | 9 | 5 | 5 | 1 | 6 | 7 | 4 | 2 | 6 | 8 | 5 | 4 | 9 | 4 | 4 | 8 | 3 | 2 | 5 | 7 | 3 | 1 | 4 | 5 | 9 |

**6,1; 6,2; 6,3; etc.**

| TWENTY FOUR TERM REPEATING SERIES | | | | | | | | | | | | | | | | | | | | | | | | | |
|---|---|---|---|---|---|---|---|---|---|---|---|---|---|---|---|---|---|---|---|---|---|---|---|---|---|
| 6 | 1 | 7 | 8 | 6 | 5 | 2 | 7 | 9 | 7 | 7 | 5 | 3 | 8 | 2 | 1 | 3 | 4 | 7 | 2 | 9 | 2 | 2 | 4 | 6 | 1 |
| 6 | 2 | 8 | 1 | 9 | 1 | 1 | 2 | 3 | 5 | 8 | 4 | 3 | 7 | 1 | 8 | 9 | 8 | 8 | 7 | 6 | 4 | 1 | 5 | 6 | 2 |
| 6 | 3 | 9 | 3 | 3 | 6 | 9 | 6 | 6 | 3 | 9 | 3 | 3 | 6 | 9 | 6 | 6 | 3 | 9 | 3 | 3 | 6 | 9 | 6 | 6 | 3 |
| 6 | 4 | 1 | 5 | 6 | 2 | 8 | 1 | 9 | 1 | 1 | 2 | 3 | 5 | 8 | 4 | 3 | 7 | 1 | 8 | 9 | 8 | 8 | 7 | 6 | 4 |
| 6 | 5 | 2 | 7 | 9 | 7 | 7 | 5 | 3 | 8 | 2 | 1 | 3 | 4 | 7 | 2 | 9 | 2 | 2 | 4 | 6 | 1 | 7 | 8 | 6 | 5 |
| 6 | 6 | 3 | 9 | 3 | 3 | 6 | 9 | 6 | 6 | 3 | 9 | 3 | 3 | 6 | 9 | 6 | 6 | 3 | 9 | 3 | 3 | 6 | 9 | 6 | 6 |
| 6 | 7 | 4 | 2 | 6 | 8 | 5 | 4 | 9 | 4 | 4 | 8 | 3 | 2 | 5 | 7 | 3 | 1 | 4 | 5 | 9 | 5 | 5 | 1 | 6 | 7 |
| 6 | 8 | 5 | 4 | 9 | 4 | 4 | 8 | 3 | 2 | 5 | 7 | 3 | 1 | 4 | 5 | 9 | 5 | 5 | 1 | 6 | 7 | 4 | 2 | 6 | 8 |
| 6 | 9 | 6 | 6 | 3 | 9 | 3 | 3 | 6 | 9 | 6 | 6 | 3 | 9 | 3 | 3 | 6 | 9 | 6 | 6 | 3 | 9 | 3 | 3 | 6 | 9 |

**7,1; 7,2; 7,3; etc.**

| TWENTY FOUR TERM REPEATING SERIES | | | | | | | | | | | | | | | | | | | | | | | | | |
|---|---|---|---|---|---|---|---|---|---|---|---|---|---|---|---|---|---|---|---|---|---|---|---|---|---|
| 7 | 1 | 8 | 9 | 8 | 8 | 7 | 6 | 4 | 1 | 5 | 6 | 2 | 8 | 1 | 9 | 1 | 1 | 2 | 3 | 5 | 8 | 4 | 3 | 7 | 1 |
| 7 | 2 | 9 | 2 | 2 | 4 | 6 | 1 | 7 | 8 | 6 | 5 | 2 | 7 | 9 | 7 | 7 | 5 | 3 | 8 | 2 | 1 | 3 | 4 | 7 | 2 |
| 7 | 3 | 1 | 4 | 5 | 9 | 5 | 5 | 1 | 6 | 7 | 4 | 2 | 6 | 8 | 5 | 4 | 9 | 4 | 4 | 8 | 3 | 2 | 5 | 7 | 3 |
| 7 | 4 | 2 | 6 | 8 | 5 | 4 | 9 | 4 | 4 | 8 | 3 | 2 | 5 | 7 | 3 | 1 | 4 | 5 | 9 | 5 | 5 | 1 | 6 | 7 | 4 |
| 7 | 5 | 3 | 8 | 2 | 1 | 3 | 4 | 7 | 2 | 9 | 2 | 2 | 4 | 6 | 1 | 7 | 8 | 6 | 5 | 2 | 7 | 9 | 7 | 7 | 5 |
| 7 | 6 | 4 | 1 | 5 | 6 | 2 | 8 | 1 | 9 | 1 | 1 | 2 | 3 | 5 | 8 | 4 | 3 | 7 | 1 | 8 | 9 | 8 | 8 | 7 | 6 |
| 7 | 7 | 5 | 3 | 8 | 2 | 1 | 3 | 4 | 7 | 2 | 9 | 2 | 2 | 4 | 6 | 1 | 7 | 8 | 6 | 5 | 2 | 7 | 9 | 7 | 7 |
| 7 | 8 | 6 | 5 | 2 | 7 | 9 | 7 | 7 | 5 | 3 | 8 | 2 | 1 | 3 | 4 | 7 | 2 | 9 | 2 | 2 | 4 | 6 | 1 | 7 | 8 |
| 7 | 9 | 7 | 7 | 5 | 3 | 8 | 2 | 1 | 3 | 4 | 7 | 2 | 9 | 2 | 2 | 4 | 6 | 1 | 7 | 8 | 6 | 5 | 2 | 7 | 9 |

8,1; 8,2; 8,3; etc.

| TWENTY FOUR TERM REPEATING SERIES | | | | | | | | | | | | | | | | | | | | | | | | | |
|---|---|---|---|---|---|---|---|---|---|---|---|---|---|---|---|---|---|---|---|---|---|---|---|---|---|
| 8 | 1 | 9 | 1 | 1 | 2 | 3 | 5 | 8 | 4 | 3 | 7 | 1 | 8 | 9 | 8 | 8 | 7 | 6 | 4 | 1 | 5 | 6 | 2 | 8 | 1 |
| 8 | 2 | 1 | 3 | 4 | 7 | 2 | 9 | 2 | 2 | 4 | 6 | 1 | 7 | 8 | 6 | 5 | 2 | 7 | 9 | 7 | 7 | 5 | 3 | 8 | 2 |
| 8 | 3 | 2 | 5 | 7 | 3 | 1 | 4 | 5 | 9 | 5 | 5 | 1 | 6 | 7 | 4 | 2 | 6 | 8 | 5 | 4 | 9 | 4 | 4 | 8 | 3 |
| 8 | 4 | 3 | 7 | 1 | 8 | 9 | 8 | 8 | 7 | 6 | 4 | 1 | 5 | 6 | 2 | 8 | 1 | 9 | 1 | 1 | 2 | 3 | 5 | 8 | 4 |
| 8 | 5 | 4 | 9 | 4 | 4 | 8 | 3 | 2 | 5 | 7 | 3 | 1 | 4 | 5 | 9 | 5 | 5 | 1 | 6 | 7 | 4 | 2 | 6 | 8 | 5 |
| 8 | 6 | 5 | 2 | 7 | 9 | 7 | 7 | 5 | 3 | 8 | 2 | 1 | 3 | 4 | 7 | 2 | 9 | 2 | 2 | 4 | 6 | 1 | 7 | 8 | 6 |
| 8 | 7 | 6 | 4 | 1 | 5 | 6 | 2 | 8 | 1 | 9 | 1 | 1 | 2 | 3 | 5 | 8 | 4 | 3 | 7 | 1 | 8 | 9 | 8 | 8 | 7 |
| 8 | 8 | 7 | 6 | 4 | 1 | 5 | 6 | 2 | 8 | 1 | 9 | 1 | 1 | 2 | 3 | 5 | 8 | 4 | 3 | 7 | 1 | 8 | 9 | 8 | 8 |
| 8 | 9 | 8 | 8 | 7 | 6 | 4 | 1 | 5 | 6 | 2 | 8 | 1 | 9 | 1 | 1 | 2 | 3 | 5 | 8 | 4 | 3 | 7 | 1 | 8 | 9 |

9,1; 9,2; 9,3; etc.

| TWENTY FOUR TERM REPEATING SERIES | | | | | | | | | | | | | | | | | | | | | | | | | |
|---|---|---|---|---|---|---|---|---|---|---|---|---|---|---|---|---|---|---|---|---|---|---|---|---|---|
| 9 | 1 | 1 | 2 | 3 | 5 | 8 | 4 | 3 | 7 | 1 | 8 | 9 | 8 | 8 | 7 | 6 | 4 | 1 | 5 | 6 | 2 | 8 | 1 | 9 | 1 |
| 9 | 2 | 2 | 4 | 6 | 1 | 7 | 8 | 6 | 5 | 2 | 7 | 9 | 7 | 7 | 5 | 3 | 8 | 2 | 1 | 3 | 4 | 7 | 2 | 9 | 2 |
| 9 | 3 | 3 | 6 | 9 | 6 | 6 | 3 | 9 | 3 | 3 | 6 | 9 | 6 | 6 | 3 | 9 | 3 | 3 | 6 | 9 | 6 | 6 | 3 | 9 | 3 |
| 9 | 4 | 4 | 8 | 3 | 2 | 5 | 7 | 3 | 1 | 4 | 5 | 9 | 5 | 5 | 1 | 6 | 7 | 4 | 2 | 6 | 8 | 5 | 4 | 9 | 4 |
| 9 | 5 | 5 | 1 | 6 | 7 | 4 | 2 | 6 | 8 | 5 | 4 | 9 | 4 | 4 | 8 | 3 | 2 | 5 | 7 | 3 | 1 | 4 | 5 | 9 | 5 |
| 9 | 6 | 6 | 3 | 9 | 3 | 3 | 6 | 9 | 6 | 6 | 3 | 9 | 3 | 3 | 6 | 9 | 6 | 6 | 3 | 9 | 3 | 3 | 6 | 9 | 6 |
| 9 | 7 | 7 | 5 | 3 | 8 | 2 | 1 | 3 | 4 | 7 | 2 | 9 | 2 | 2 | 4 | 6 | 1 | 7 | 8 | 6 | 5 | 2 | 7 | 9 | 7 |
| 9 | 8 | 8 | 7 | 6 | 4 | 1 | 5 | 6 | 2 | 8 | 1 | 9 | 1 | 1 | 2 | 3 | 5 | 8 | 4 | 3 | 7 | 1 | 8 | 9 | 8 |
| 9 | 9 | 9 | 9 | 9 | 9 | 9 | 9 | 9 | 9 | 9 | 9 | 9 | 9 | 9 | 9 | 9 | 9 | 9 | 9 | 9 | 9 | 9 | 9 | 9 | 9 |

**Results:**

- Each series (with the exception of the nine series that start with combinations of 3, 6 and 9), has a 24 term cycle, meaning that they repeat every 24 terms, .

- The series with combinations of 3, 6, and 9 are: 3,3; 3,6; 3,9; 6,3; 6,6; 6,9; 9,3; 9,6; 9,9. The numbers in these series are shown italicized in the tables. and we notice that there are only three digits used, in this eight series cycle.

- The 9,9 series is special as all of the terms are SDQ 9, and these numbers are shown underlined in the table.

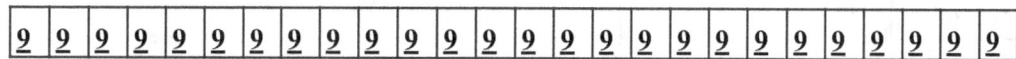

| 9 | 9 | 9 | 9 | 9 | 9 | 9 | 9 | 9 | 9 | 9 | 9 | 9 | 9 | 9 | 9 | 9 | 9 | 9 | 9 | 9 | 9 | 9 | 9 |
|---|---|---|---|---|---|---|---|---|---|---|---|---|---|---|---|---|---|---|---|---|---|---|---|

- The eight other combinations of 3, 6, and 9 each has a cycle of 8 terms. For example, the 6, 3 series is here reproduced:

| 6 | 3 | 9 | 3 | 3 | 6 | 9 | 6 | 6 | 3 | 9 | 3 | 3 | 6 | 9 | 6 | 6 | 3 | 9 | 3 | 3 | 6 | 9 | 6 | 6 | 3 |
|---|---|---|---|---|---|---|---|---|---|---|---|---|---|---|---|---|---|---|---|---|---|---|---|---|---|

In each of these eight series, each cycle contains nine 3's, nine 6's, and six 9's, for a total of 135, which also has an SDQ of 9, i.e. 135 ==> 9.

- Every 24 term cycle series contains two of each digit, two of the digits, however, occur five times. The two digits that occur five times, add to 9. The sum of the terms in the 24 term cycle series add to 117 ==> 9. This can be seen by noting that every digit occurs twice, which adds to 90, and there are two digits, that total 9, that occur an additional three times, or 27. Thus, 90 + 27 = 117.

-In each series, where the first term is repeated, so is the thirteenth term, and the first term and the thirteenth term add to 9.

-For any given table, numbers in the 2nd column and their corresponding numbers in the 23rd column, all have the same total.

-In any given table, the 5th column contains only three different numbers, as does the 21st column, but the numbers are not in the same order.

**Double Starting Number Series**

The next table summarizes the results for the double number series. Note that as before, there are 2 series that have 8 term cycles, and one that is all 9's. As before, the 8 term cycles have their numbers italicized, and the all 9 series is underlined.

| TWENTY FOUR TERM REPEATING SERIES | | | | | | | | | | | | | | | | | | | | | | | | | |
|---|---|---|---|---|---|---|---|---|---|---|---|---|---|---|---|---|---|---|---|---|---|---|---|---|---|
| 1 | 1 | 2 | 3 | 5 | 8 | 4 | 3 | 7 | 1 | 8 | 9 | 8 | 8 | 7 | 6 | 4 | 1 | 5 | 6 | 2 | 8 | 1 | 9 | 1 | 1 |
| 2 | 2 | 4 | 6 | 1 | 7 | 8 | 6 | 5 | 2 | 7 | 9 | 7 | 7 | 5 | 3 | 8 | 2 | 1 | 3 | 4 | 7 | 2 | 9 | 2 | 2 |
| *3* | *3* | *6* | *9* | *6* | *6* | *3* | *9* | *3* | *3* | *6* | *9* | *6* | *6* | *3* | *9* | *3* | *3* | *6* | *9* | *6* | *6* | *3* | *9* | *3* | *3* |
| 4 | 4 | 8 | 3 | 2 | 5 | 7 | 3 | 1 | 4 | 5 | 9 | 5 | 5 | 1 | 6 | 7 | 4 | 2 | 6 | 8 | 5 | 4 | 9 | 4 | 4 |
| 5 | 5 | 1 | 6 | 7 | 4 | 2 | 6 | 8 | 5 | 4 | 9 | 4 | 4 | 8 | 3 | 2 | 5 | 7 | 3 | 1 | 4 | 5 | 9 | 5 | 5 |
| *6* | *6* | *3* | *9* | *3* | *3* | *6* | *9* | *6* | *6* | *3* | *9* | *3* | *3* | *6* | *9* | *6* | *6* | *3* | *9* | *3* | *3* | *6* | *9* | *6* | *6* |
| 7 | 7 | 5 | 3 | 8 | 2 | 1 | 3 | 4 | 7 | 2 | 9 | 2 | 2 | 4 | 6 | 1 | 7 | 8 | 6 | 5 | 2 | 7 | 9 | 7 | 7 |
| 8 | 8 | 7 | 6 | 4 | 1 | 5 | 6 | 2 | 8 | 1 | 9 | 1 | 1 | 2 | 3 | 5 | 8 | 4 | 3 | 7 | 1 | 8 | 9 | 8 | 8 |
| <u>9</u> | <u>9</u> | <u>9</u> | <u>9</u> | <u>9</u> | <u>9</u> | <u>9</u> | <u>9</u> | <u>9</u> | <u>9</u> | <u>9</u> | <u>9</u> | <u>9</u> | <u>9</u> | <u>9</u> | <u>9</u> | <u>9</u> | <u>9</u> | <u>9</u> | <u>9</u> | <u>9</u> | <u>9</u> | <u>9</u> | <u>9</u> | <u>9</u> | <u>9</u> |

This is like the 9 table, with all columns moved over one column to the right.

## SDQ AND FIBONACCI SQUARES

The regular Fibonacci series is given by:
1, 1, 2, 3, 5, 8, 13, 21, 34, 55, 89, 144, 233, 377, 610, 987, 1,597, 2,584, 4,181, 6,765, 10,946, 17,711, 28,657, 46,368; 75,025, 121,393, 196,418. The Fibonacci series of squares are the value of the square of each term of the series, thus, 1, 1, 4, 9, 25, 64, 169, 441, 1156, 3025, 7921, 20736; 54289, 142129, 372100, 974169, 2550409, ...
The ratio of a term to its preceding term in the regular Fibonacci Series yields the Golden Proportion 1.618+ called $\phi$.

The ratio for the Fibonacci Squares is 2550409/974169 = 2.618+ or $\phi^2$.

If we now take the SDQ of each term we get the following series of SDQ squares: 1, 1, 4, 9, 7, 1, 7, 9, 4, 1, 1, 9; 1, 1, 4, 9, 7, 1, 7, 9, 4, 1, 1, 9; The Fibonacci squares are a 12 term repeating series in the SDQ. Like the regular Fibonacci series, the terms are; male, male, female, male, male, female, ... . This is the same as the regular series since squaring does not change the sex of the number. Odd numbers remains odd, and even numbers remains even. The sum of the first twelve terms of the square Fibonacci series is 33,552 => 18 => 9. The sum of the first twelve terms of the SDQ repeating series is: 54 => 9.

Another equivalent method of getting the final SDQ of the square Fibonacci series, is to take the SDQ of each term, square this SDQ, and then perform the SDQ again. Using SDQ of the original series, this series becomes: 1, 1, 2, 3, 5, 8, 4, 3, 7, 1, 8, 9, 8, 8, 7, 6, 4, 1, 5, 6, 2, 8, 1, 9; 1, 1, 2, ...
Now, square each term, we arrive at the following: 1, 1, 4, 9, 7, 1, 7, 9, 4, 1, 1, 9; 1, 1, 4, 9, 7, ... . This series has a cycle of 12, half that of the of the original series whose cycle is 24. The sum of this 12 term cycle is 54 => 9.

The latter method is preferred to determine the value of a series whose individual terms become excessively large when raised to higher order. We note however, that the former method using the original series terms, must be used to determine the pattern of the primary male and the primary female.

The next feature noticeable is that this series is comprised of only the four digits; **1**, **4**, **7**, and **9**.

Another feature is that the SDQ square Fibonacci pattern 1, 1, 4, 9, 7, 1, 7, 9, 4, 1, 1, 9; can be analyzed by looking at the pattern of 9's. There are three sets of four digits each that end with 9, namely: 1, 1, 4, 9; 7, 1, 7, 9; and 4, 1, 1, 9. **Each of these triplets, has a sum whose SDQ is 6. The SDQ of the triplet is 18 => 9.**

### SDQ AND THE SQUARE FIBONACCI-LIKE SERIES

| TWELVE TERM REPEATING SERIES | | | | | | | | | | | | | | | | | | | | | | | | | |
|---|---|---|---|---|---|---|---|---|---|---|---|---|---|---|---|---|---|---|---|---|---|---|---|---|---|
| 1 | 1 | 4 | 9 | 7 | 1 | 7 | 9 | 4 | 1 | 1 | 9 | 1 | 1 | 4 | 9 | 7 | 1 | 7 | 9 | 4 | 1 | 1 | 9 | 1 | 1 |
| 1 | 2 | 9 | 7 | 1 | 7 | 9 | 4 | 1 | 1 | 9 | 1 | 1 | 4 | 9 | 7 | 1 | 7 | 9 | 4 | 1 | 1 | 9 | 1 | 1 | 4 |
| 1 | 3 | 7 | 4 | 4 | 9 | 4 | 4 | 7 | 9 | 1 | 4 | 1 | 9 | 7 | 4 | 4 | 9 | 4 | 4 | 7 | 9 | 1 | 4 | 1 | 9 |
| 1 | 4 | 7 | 9 | 7 | 7 | 1 | 9 | 4 | 7 | 4 | 9 | 1 | 7 | 7 | 9 | 7 | 7 | 1 | 9 | 4 | 7 | 4 | 9 | 1 | 7 |
| 1 | 5 | 9 | 4 | 1 | 1 | 9 | 1 | 1 | 4 | 9 | 7 | 1 | 7 | 9 | 4 | 1 | 1 | 9 | 1 | 1 | 4 | 9 | 7 | 1 | 7 |
| 1 | 6 | 4 | 7 | 4 | 9 | 1 | 7 | 7 | 9 | 7 | 7 | 1 | 9 | 4 | 7 | 4 | 9 | 1 | 7 | 7 | 9 | 7 | 7 | 1 | 9 |
| 1 | 7 | 1 | 9 | 7 | 4 | 4 | 9 | 4 | 4 | 7 | 9 | 1 | 4 | 1 | 9 | 7 | 4 | 4 | 9 | 4 | 4 | 7 | 9 | 1 | 4 |
| 1 | 8 | 9 | 1 | 1 | 4 | 9 | 7 | 1 | 7 | 9 | 4 | 1 | 1 | 9 | 1 | 1 | 4 | 9 | 7 | 1 | 7 | 9 | 4 | 1 | 1 |
| 1 | 9 | 1 | 1 | 4 | 9 | 7 | 1 | 7 | 9 | 4 | 1 | 1 | 9 | 1 | 1 | 4 | 9 | 7 | 1 | 7 | 9 | 4 | 1 | 1 | 9 |

We notice the twelve term repeating series in the SDQ, but must mention that other than the 1,1 series, the repeating series starts after the first cycle, since the first two starting numbers in each series is not squared. **The cycle starts with the underlined numbers.**

| 1 | 2 | 9 | 7 | 1 | 7 | 9 | 4 | 1 | 1 | 9 | 1 | 1 | 4 | 9 | 7 | 1 | 7 | 9 | 4 | 1 | 1 | 9 | 1 | 1 | 4 |
|---|---|---|---|---|---|---|---|---|---|---|---|---|---|---|---|---|---|---|---|---|---|---|---|---|---|

## SDQ AND FIBONACCI CUBES

The regular Fibonacci series is given by:

1, 1, 2, 3, 5, 8, 13, 21, 34, 55, 89, 144, 233, 377, 610, 987, 1,597, 2,584, 4,181, 6,765, 10,946, 17,711, 28,657, 46,368; 75,025, 121,393, 196,418, ...

The cubed Fibonacci series is:

1, 1, 8, 27, 125, 512, 2197, 9261; 39304, 166375, 704969, 1481544, 12649337, 53582633, .... . This still exhibits the male, male, female pattern.

The ratio of a term to its preceding term is 53582633/12649337 = 4.2360+ or $\phi^3$.

The SDQ of the cubed Fibonacci series is: 1, 1, 8, 9, 8, 8, 1, 9; 1, 1, 8, 9, 8, 8,

This is an 8 term repeating series; 1, 1, 8, 9, 8, 8, 1, 9, whose sum is 45 => 9. The sum of the first eight terms is 12,132 => 9.

The next feature is that this entire SDQ series is composed of only the three digits; **1, 8, and 9.**

Another feature is that this pattern can be analyzed by looking at the pattern of 9's as in the Fibonacci square SDQ series. There are two sets of four digits each that end in 9, namely: 1, 1, 8, 9 and 8, 8, 1, 9. **These doublets have the sum whose SDQ is 1 and 8, respectively, and thus the doublet total is 9.**

| EIGHT TERM REPEATING SERIES | | | | | | | | | | | | | | | | | | | | | | | | | |
|---|---|---|---|---|---|---|---|---|---|---|---|---|---|---|---|---|---|---|---|---|---|---|---|---|---|
| 1 | 1 | 8 | 9 | 8 | 8 | 1 | 9 | 1 | 1 | 8 | 9 | 8 | 8 | 1 | 9 | 1 | 1 | 8 | 9 | 8 | 8 | 1 | 9 | 1 | 1 |
| 1 | 2 | 9 | 8 | | | | | | | | | | | | | | | | | | | | | | |
| 1 | 3 | | | | | | | | | | | | | | | | | | | | | | | | |
| 1 | 4 | | | | | | | | | | | | | | | | | | | | | | | | |
| 1 | 5 | | | | | | | | | | | | | | | | | | | | | | | | |
| 1 | 6 | | | | | | | | | | | | | | | | | | | | | | | | |
| 1 | 7 | | | | | | | | | | | | | | | | | | | | | | | | |
| 1 | 8 | | | | | | | | | | | | | | | | | | | | | | | | |
| 1 | 9 | | | | | | | | | | | | | | | | | | | | | | | | |

Other than the first two starting numbers, all SDQ digits are either 1, 8, or 9.

## SDQ AND FIBONACCI TO THE FOURTH POWER

The regular Fibonacci series is given by:

1, 1, 2, 3, 5, 8, 13, 21, 34, 55, 89, 144, 233, 377, 610, 987, 1,597, 2,584, 4,181, 6,765, 10,946, 17,711, 28,657, 46,368; 75,025, 121,393, 196,418

The fourth power Fibonacci series is:

1, 1, 16, 81, 625, 4096, 28561, 194481, 1336336, 9150625, 62742241, 429981696, 2947295521, ... . Again we still have the male, male, female pattern.

The ratio of a term to its preceding term is 2947295521/429981696 = 6.8545+ or $\phi^4$.

The SDQ are:1, 1, 7, 9, 4, 1, 4, 9, 7, 1, 1, 9; a triplet of four digits each, each triplet has the sum of 18, and the total sum us 54 ==> 9.

**We see the remarkable relationship between the square and the quartic SDQ Fibonacci series, where the 4 and the 7 change places.**
1, 1, 4, 9, 7, 1, 7, 9, 4, 1, 1, 9; (square)
1, 1, 7, 9, 4, 1, 4, 9, 7, 1, 1, 9; (quartic).

**SQUARED NUMBERS**

All squared numbers have additional properties. Consider a sequence starting with the numbers sequential numbers, 3 and 4. Thus, 3, 4, 5, 6, 7, 8, .... The square of each term is: 9, 16, 25, 36, 49, 64, 81, 100, 121, 144, 169, 196, 225, 256, 289, 324, 361, 400, 441, 484, 529, 576, 625, 676, 729, 784, 841, 900, 961, 1024, etc.

The period is 9, and the sum of the SDQ is 501 => 6.

This series of successive squares has the SDQ values: 9, 7, 7, 9, 4, 1, 9, 1, 4; 9, 7, 7, 9, 4, 1, 9, 1, 4; ...

Notice that only the digits 1, 4, 7, and 9 are used. [rem: find a use for this if possible, 1+4+7+9=21=3.]

Applying the pattern of 9, established by the SDQ of the squares, to the original sequence, namely: 3, 4, 5, 6, 7, 8, 9, 10, 11, 12, 13, 14, 15, 16, 17, 18, 19, 20, 21, 22, 23, 24, 25, 26, 27, 28, 29, and adding each 9 pattern sum, yields the following; 63, 144, 225, and so on for each succeeding pattern sum. The SDQ of each pattern becomes: 63 => 9, 144 => 9, 225 => 9. This property is only apparent from the pattern formed by the SDQ of the squares.

In the original series, every third term, has an SDQ of 9, and the SDQ of it's square is also 9, which makes sense, since any SDQ of 9, when squared still yields an SDQ of 9.


**TRIBONACCI SERIES: Tri-Fi Series**

In the "usual" Fibonacci series, each term is the sum of the two prior terms, a "Bi-Fi" series. Now consider a series where the next term is the sum of the three prior terms, a "Tri-Fi" series.

The first terms could be: 1  2  3  followed by 6  which is the sum of the three prior terms 1  2  and 3.

The Tri-Fibo series would be as follows:

1 2 3 6 11 20 37 68 125 230 423 778 1,431 2,632 4,841 8,904 16,377 30,122 55,403  101,902  187,427  344,732  634,061  1,166,220  2,145,013  3,945,294 7,256,527 13,346,834 24,548,655 45,152,016 83,047,505 152,748,176 280,947,697 516,743,378  950,439,251  1,748,130,326  3,215,312,955  5,913,882,532 10,877,325,813; 20,006,521,300 …

**The SDQ of this Tri-Fibo series is:**

1 2 3 6 2 2 1 5 8 5 9 4 9 4 8 3 6 8 8 4 2 5 2 9 7 9 7 5 3 6 5 5 7 8 2 8
9 1 9; 1 2 3

This is a 39 term repeating series.
The ratio of a term to its prior term is 1.839 and the ratio of a term to its
following term is 0.544.

**Quadrinacci Series: Quadri-Fibo Series**

This series is characterized by a term being the sum of its four prior terms. Thus,

1 2 3 4 10 19 36 69 134 258 497 958 1,847 ...

**The SDQ of this Quadri-Fibo Series is:**

1 2 3 4 1 1 9 6 8 6 2 4 2 5 4 6 8 5 5 6 6 4 3 1 5 4 4 5 9 4 4 4 3 6 8 3
2 1 5 2 1 9 8 2 2 3 6 4 6 1 8 1 7 8 6 4 7 7 6 6 8 9 2 7 8 8 7 3 8 8 8 9
6 4 9 1 2 7; 1 2 3 4 1 1 9 6 8 6 2 4 2 5 4 6 8 5 5 6 6 4 3 1 5 4 4 5 9
4 4 4 3 6 8 3 2 1 5 2 1 9 8 2 2 3 6 4 6 1 8 1 7 8 6 4 7 7 6 6 8 9 2 7 8
8 7 3 8 8 8 9 6 4 9 1 2 7; 1 2 3 4 1 1 9 6 ...

This is a 78 terms repeating series, with the term ratios of 1.928 and 0.518.

**Pentanacci Series: Penta-Fibo Series**

This series is characterized by a term being the sum of its five prior terms. Thus,

1 2 3 4 5 15 29 56 109 214 423 831 1633 3210 6311 12408 24393 …

**The SDQ of this Penta-Fibo Series is:**
1 2 3 4 5 6 2 2 1 7 9 3 4 6 2 6 3 3 2 7 3 9 6 9 7 7 2 4 2 4 1 4 6 8 5 6
2 9 3 7 9 3 4 8 4 1 2 1 7 6 8 6 1 1 4 2 5 4 7 4 4 6 7 1 4 4 4 2 6 2 9 5
6 1 5 8 7 9 3 5 5 2 6 3 3 1 6 1 5 7 2 3 9 8 2 6 1 8 7 6 1 5 9 1 4 2 3 1
2 3 2 2 1 1 9 6 1 9 8 6 3 9 8 7 6 6 9 9 1 4 2 7 5 1 1 7 3 8 2 3 5 3 3 7
3 3 1 8 4 1 8 4 7 6 8 6 4 4 1 5 2 7 1 7 4 3 4 1 1 4 4 5 6 2 3 2 9 4 2 2
1 9 9 5 8 5 9 9 9 4 9 4 8 7 5 6 3 2 5 3 1 5 7 3 1 8 6 7 7 2 3 7 8 9 2 2
1 4 9 9 7 3 5 6 3 6 5 7 9 3 3 9 4 1 2 1 8 7 1 1 9 8 8 9 8 6 3 7 6 3 7 8
4 1 5 7 7 6 8 6 7 7 7 8 8 1 4 1 4 9 1 1 7 4 4 8 6 2 6 8 3 7 8 5 4 9 6 5
2 8 3 6 6 7 3 7 2 7 8 9 6 5 8 9 1 2 7 9 1 2 3 4; 1 2 3 4 5 6 2 2

312 term period!!!

To summarize:

**The SDQ on their frequencies for these series are:**

|   | Bi-Fi | Tri-Fi | Quadri-Fi | Penta-Fi |
|---|---|---|---|---|
| 1 | 5 | 3 | 9 | 42 |
| 2 | 2 | 6 | 9 | 36 |
| 3 | 2 | 3 | 6 | 36 |
| 4 | 2 | 3 | 12 | 36 |
| 5 | 2 | 6 | 6 | 24 |
| 6 | 2 | 3 | 12 | 36 |
| 7 | 2 | 3 | 6 | 39? |
| 8 | 5 | 6 | 12 | 30 |
| 9 | 2 | 6 | 6 | 33? |
| Total 24 | | 39 | 78 | 312 |
| SDQ 6 | | 3 | 6 | 6 |

Ratios:

| Upper | 1.618 | 1.839 | 1.928 | 1.966 | 1.984 | 1.992 |
|---|---|---|---|---|---|---|

2.000

| Lower | 0.618 | 0.544 | 0.518 | 0.509 | 0.504 | 0.502 |
| 0.500 | | | | | | |
| U + L | 2.236 | 2.383 | 2.446 | 2.475 | 2.488 | 2.494 |
| 2.500 | | | | | | |
| U – L | 1.000 | 1.295 | 1.410 | 1.457 | 1.480 | 1.490 |
| 1.500 | | | | | | |
| U/L | 2.618 | 3.380 | 3.722 | 3.862 | 3.936 | 3.968 |
| 4.000 | | | | | | |

**It appears that the upper number approaches 2.0 and the lower approaches 0.5. Or to put it another way: the upper approaches 2 and the lower $2^{-1}$.**

Quick test of a Fibo series. Take the ratio and inverse ratio of any two large consecutive terms and determine which Fibo series, if any, it relates to.

**Fi-Fi**

1 2 3 4 5 15 29 56 109 214 423 831 1633 3210 6311 12408 24393

**Si-Fi**

1 2 3 4 5 6 21 41 80 157 310 615 1224 2427 4813 9546 18935 37560

**Sept-Fi**

1 2 3 4 5 6 7 28 55 108 213 422 839 1672 3337 6646 13237 26366

**Product Bi-Fibo**
Here $F_{n+2} = F_{n+1}F_n$ where n = 1, 2, … [N.B. Cannot start with 0.]
Each term is the product of the two prior terms:
If either of the two digits is a 3, 6 or 9 it will be all 9 after the third term. If one of the digits is a 9, it will be all 9 after the second digit.
We show the SDQ for each term. The first number is $F_1$ the second $F_2$

$F_1 = 1, F_2 = 2, 3, …, 9$

1 **2** 2 4 8 5 4 2 8 7 2 5 1 5 5 7 8 2 7 5 8 4 5 2; 1 2 2 4

113

24 term Repeat Block like the usual Fibonacci series.   SDQ(RB) = 109 => 1

**1 3** 3 9 9 9 9 9 9 9 9 9                                          all 9

**1 4** 4 7 1 7 7 4; 1 4 4 7 1 7 7 4

    **8** term Repeat Block,                              SDQ(RB) = 35 => 8

**1 5** 5 7 8 2 7 5 8 4 5 2 1 2 2 4 8 5 4 2 8 7 2 5; 1 5 5 7

    **24** term Repeat Block,                            SDQ(RB) = 109 => 1

**1 6** 6 9 9 9 9 9 9 9 9                                          all 9

**1 7** 7 4 1 4 4 7; 1 7 7 4 1

    **8** term Repeat Block,                              SDQ(RB) = 35 => 8

**1 8** 8; 1 8 8 1 8 8 1

    **3** term Repeat Block,                              SDQ(RB) = 17 => 8

**1 9** 9 9 9 9 9 9 9 9                                          all 9

$F_1 = 2, F_2 = 1, 2, 3, …, 9$

**2 1** 2 2 4 8 5 4 2 8 7 2 5 1 5 5 7 8 2 7 5 8 4 5; 2 1 2 **2**

    **24** term Repeat Block,                            SDQ(RB) = 109 => 1

**2 2** 4 8 5 4 2 8 7 2 5 1 5 5 7 8 2 7 5 8 4 5 2 1; 2 2 4 8

    **24** term Repeat Block,                            SDQ(RB) = 109 => 1

**2 3** 6 9 9 9 9 9 9 9 9                                          all 9

**2 4** 8 5 4 2 8 7 2 5 1 5 5 7 8 2 7 5 8 4 5 2 1 2; 2 4 8 5

    **24** term Repeat Block,                            SDQ(RB) = 109 => 1

**2 5** 1 5 5 7 8 2 7 5 8 4 5 2 1 2 2 4 8 5 4 2 8 7; 2 5 1 5

    **24** term Repeat Block,                            SDQ(RB) = 109 => 1

**2 6** 3 9 9 9 9 9 9 9 9                                          all 9

**2 7** 5 8 4 5 2 1 2 2 4 8 5 4 2 8 7 2 5 1 5 5 7 8; 2 7 5 8

    **24** term Repeat Block,                            SDQ(RB) = 109 => 1

**2 8** 7 2 5 1 5 5 7 8 2 7 5 8 4 5 2 1 2 2 4 8 5 4; 2 8 7 2

    **24** term Repeat Block,                            SDQ(RB) = 109 => 1

**2 9** 9 9 9 9 9 9 9 9 9                                          all 9

$F_1 = 3, F_2 = 1, 2, 3, …, 9$

**3 1** 3 3 9 9 9 9 9 9 9                                          all 9

**3 2** 6 3 9 9 9 9 9 9 9                                          all 9

**3 3** 9 9 9 9 9 9 9 9 9                                          all 9

**3 4** 3 3 9 9 9 9 9 9 9                                          all 9

**3 5** 6 3 9 9 9 9 9 9 9                                          all 9

```
3 6 9 9 9 9 9 9 9 9 9 9                                    all 9
3 7 3 3 9 9 9 9 9 9 9 9                                    all 9
3 8 6 3 9 9 9 9 9 9 9 9                                    all 9
3 9 9 9 9 9 9 9 9 9 9 9                                    all 9
```

$F_1 = 4, F_2 = 1, 2, 3, ..., 9$

**4 1** 4 4 7 1 7 7; 4 1 4 4 7 1 7 7 4

     **8** term Repeat Block,                                 SDQ(RB) = 35 => 8

**4 2** 8 7 2 5 1 5 5 7 8 2 7 5 8 4 5 2 1 2 2 4 8 5; 4 2 8 7

     **24** term Repeat Block,                                SDQ(RB) = 109 => 10=> 1

**4 3** 3 9 9 9 9 9 9 9 9                                  all 9

**4 4** 7 1 7 7 4 1; 4 4 7 1 7 7 4 1

     8 term Repeat Block,                                   SDQ(RB) = 35 => 8

**4 5** 2 1 2 2 4 8 5 4 2 8 7 2 5 1 5 5 7 8 2 7 5 8; 4 5 2 1 2 2 4

     24 term Repeat Block,                                  SDQ(RB) = 109 => 1

$F_1 = 5, F_2 = 1, 2, 3, ..., 9$

5 1 5 5 7 8 2 7 5 8 4 5 2 1 2 2 4 8 5 4 2 8 7 2; 5 1 5 5

     24 term Repeat Block,                                  SDQ(RB) = 109 => 1

5 2 1 2 2 4 8 5 4 2 8 7 2 5 1 5 5 7 8 2 7 5 8 4; 5 2 1 2 2

**Patterns**

It appears that: Every 24 term Repeat Block has an SDQ of 1 and the same frequencies of digits:
two 1's, six 2's, three 4's, six 5's, three 7's, and four 8's, and no 3, 6 or 9.

All the single digit combinations replicate the totality of any two starting numbers, which are the same as their SDQ counterparts.

Any first and second term with an SDQ of 3, 6 or 9 will yield all 9's.

1 2 3 6 9 9 9 9 9 9 9 9 9

2 3 4 6 9 9 9 9 9 9 9 9 9

1 3 5 6 9 9 9 9 9 9 9 9 9

1 3 7 3 9 9 9 9 9 9 9 9 9

1 4 6 6 9 9 9 9 9 9 9 9 9

2 4 8 1 5 4 2 4 5 4 8 7 8 7 5 1 8 4 5 7 5 4 5 1 2 1; 2 4 8 1     **26** term period

2 5 7 7 2 8 4 1 5 2 1 1 2 2 4 7 2 2 1 4 8 5 7 1 8 2 7 4 2 2 7 1 5 8 4 7 8 8 7 7 5 2 7 7 8 5 1 4 2 8 1 7; 2 5 7 7 2     **52** term period

2 5 8 8 5 5 2 5 5 5 8 2 8; 2 5 8 8 5 5 2 5 5 5 8 2 8 2 5     **13** term period

4 5 7 5 4 5 1 2 1 2 4 8 1 5 4 2 4 5 4 8 7 8 7 5 1 8 4 5 7 5 4 5 1 2 1 2 4 8 1 5 4 2 4 5 4 8 7 8 7 5 1 8; 4 5 7 5 4     **52** term period

4 7 8 8 7 7 5 2 7 7 8 5 1 4 2 8 1 7 2 5 7 7 2 8 4 1 5 2 1 1 2 2 4 7 2 2 1 4 8 5 7 1 8 2 7 4 2 2 7 1 5 8; 4 7 8 8 7 7 5 2 7 7 8 5     **52** term period

4 6 8 3 9 9 9 9 9 9

4 6 7 6 9 9 9 9 9

4 8 2 1 7 5 8 1 4 5 2 4 4 5 8 7 1 2 5 1 1 5 5 7 4 5 5 1 7 8 2 4 1 8 5 4 7 5 5 4 1 2 8 7 4 8 8 4 4 2 5 4; 4 8 2 1 7 5     **52** term period

It appears that if any of the initial series digits is a 3, 6, or 9, at some point in the series all of the remaining digits will be 9's. This is due to the fact that once a 9 occurs in a product that 9 will reflect itself in every remaining term.

Examples:

2 3 8 3 9 9 9 9 9
2 6 8 6 9 9 9 9 9
2 2 6 6 9 9 9 9 9

## Product Four-Fi

1 2 4 5 4 7 2 1 2 1 4 8 1 5 7 1 8 1 2 7 4 2 4 8 4 4 8 7 5 4 4 2 7 4 8 7
2 7 1 8 4 8 4 7 5 4 2 1 4 5 4 8 1 7 8 7 5 7 7 5 1 2 7 7 8 1 5 1 4 2 8 1
1 7 2 5 7 4 1 5 1 2 1 1 2 4 8 1 1 5 4 2 4 7 8 7 2 1 4 2 7 2 4 4 8 4 8 7
1 8 7 5 1 1 8 4 5 7 4 2 1 2 7 1 5 7 2 7 4 5 1 5 1 7 8 1 2 4 1 8 1 5 4 7
5 7 8 7 7 8 4 2 7 7 5 4 8 4 1 2 1 8 7 4 8 1 8 4 4 2 4 1 5 4 8 7 4 5 4 2
7 1 2 1 5 1 1 5 7 8 1 1 2 7 5 7 4 8 4 5 1 7 5 4 5 7 7 8 7 8 4 1 8 4 2 1
1 8 7 2 4 7 5 1 5 4; 1 2 4 5 4 7                     **226** term period

2 5 7 8 2 2 8 4 2 2 2 5 4 8 5 8 2 1 8 2 5 8 1 8 5 5 2 4 2 8 2 2 1 5 2 2
2 4 5 8 5 8 7 8 8 2 5 1 8 8 5 5 7 5 2 8 2 7 8 5 2 2 7 5 5 8 5 1 2 8 8 2
4 8 8 8 5 4 2 5 2 8 7 2 8 5 2 7 2 5 5 8 4 8 2 8 8 7 5 8 8 8 4 5 2 5 2 1
2 2 8 5 7 2 2 5 5 1 5 8 2 8 1 2 5 8 8 1 5 5 2 5 7 8 2 2 8 4 2 2 2 5 4 8
5 8 2 1 8 2 5 8 1 8 5 5 2 4 2 8 2 2 1 5 2 2 2 4 5 8 5 8 7 8 8 2 5 1 8 8
5 5 7 5 2 8 2 7 8 5 2 2 7 5 5 8 5 1 2 8 8 2 4 8 8 8 5 4 2 5 2 8 7 2 8 5
2 7 2 5 5 8 4 8 2 8 8 7 5 8 8 8 4 5 2 5 2 1 2 2 8 5 7 2 2 5 5 1 5 8 2 8
1 2 5 8 8 1 5 5; 2 5 7 8 2 2 8                       **260** term period

1 2 7 8 4 7 2 7 5 4 1 5 1 2 1 1 2 4 8 1 1 5 4 2 4 7 8 7 2 1 4 2 7 2 4 4
8 4 8 7 1 8 7 5 1 1 8 4 5 7 4 2 1 2 7 1 5 7 2 7 4 5 1 5 1 7 8 1 2 4 1 8
1 5 4 7 5 7 8 7 7 8 4 2 7 7 5 4 8 4 1 2 1 8 7 4 8 1 8 4 4 2 4 2 1 7 2 1
5 7 7 2 4 5 1 4 8 7 8 1 7 5 1 8 1 4 5 7 5 7 1 2 7 8 4 7 2 7 5 4 1 5 1 2
1 1 2 4 8 1 1 5 4 2 4 7 8 7 2 1 4 2 7 2 4 4 8 4 8 7 1 8 7 5 1 1 8 4 5 7
4 2 1 2 7 1 5 7 2 7 4 5 1 5 1 7 8 1 2 4 1 8 1 5 4 7 5 7 8 7 7 8 4 2 7 7

117

5 4 8 4 1 2 1 8 7 4 8 1 8 4 4 2 4 2 1 7 2 1 5 7 7 2 4 5 1 4 8 7 8 1 7 5
1 8 1 4 5 7 5 7; 1 2 7 8 4 7 2 7                          **260** term period

1 1 1 5 5 7 4 7 8 2 7 1 4 2 2 7 4 4 8 5 1 7 1 8 2 4 1 1 8 5 4 7 4 2 8 7
7 1 5 2 7 7 4 5 8 4 1 7 8 8 7 4 1 8 8 4 4 7 5 2 1 7 7 8 5 7 7 7 5 5 1 4
1 2 8 1 7 4 8 8 1 4 4 2 5 7 1 7 2 8 4 7 7 2 5 4 1 4 8 2 1 1 7 5 8 1 1 4
5 2 4 7 1 2 2 1 4 7 2 2 4 4 1 5 8 7 1 1 2 5; 1 1 1 5 5 7
                                                         **44** term period

1 1 1 2 2 4 7 4 8 5 4 1 7 5 5 4 7 7 8 2 1 4 1 8 5 7 1 1 8 2 7 4 7 5 8 4
4 1 2 5 4 4 7 2 8 7 1 4 8 8 4 7 1 8 8 7 7 4 2 5 1 4 4 8 2 4 4 4 2 2 1 7
1 5 8 1 4 7 8 8 1 7 7 5 2 4 1 4 5 8 7 4 4 5 2 7 1 7 8 5 1 1 4 2 8 1 1 7
2 5 7 4 1 5 5 1 7 4 5 5 7 7 1 2 8 4 1 1 5 2; 1 1 1 2 2 4 7
                                                         **44** term period

## Fibonacci Primes [FAR]

2  3  5  13  89  233  1597  28657  514229

## SDQ of Fibonacci Primes
**2 3 5 4 8 8 4 1 5**
## General Fibo-like Plus-Minus Series
This series is given by;
$F_{n+2} = F_{n+1} + F_n \pm a$ where a is any number and n = 1, 2, 3…
Here $F_1 = 1$, and $F_2 = 2$. (Fibonacci-like)
$F_1 = 1$, $F_2 = 2$, $F_3 = 3 \pm a$, $F_4 = 5 \pm 2a$, $F_5 = 8 \pm 4a$, $F_6 = 13 \pm 7a$, $F_7 = 21 \pm 12a$, $F_8 = 34 \pm 20a$,
$F_9 = 55 \pm 33a$, $F_{10} = 89 \pm 54a$, $F_{11} = 144 \pm 88a$, $F_{12} = 233 \pm 143a$, $F_{13} = 377 \pm 232a$, $F_{14} = 610 \pm 376a$,
$F_{15} = 987 \pm 609a$, $F_{16} = 1,597 \pm 986a$, $F_{17} = 2,584 \pm 1,596a$, $F_{18} = 4,181 \pm 2,583a$, $F_{19} = 6,765 \pm 4,180a$,
$F_{20} = 10,946 \pm 6,764a$, $F_{21} = 17,711 \pm 10,945a$, $F_{22} = 28,657 \pm 17,710a$, $F_{23} = 46,368 \pm 28,656a$,
$F_{24} = 75,025 \pm 46,367a$, $F_{25} = 121,393 \pm 75,024a$, $F_{26} = 196,418 \pm 121,392a$, $F_{27} = 317,811 \pm 196,417a$,

We note that $SDQ(F_{n+2}) = SDQ(F_{n+1} + F_n \pm a) = SDQ(F_{n+1}) + SDQ(F_n) \pm SDQ(a)$
This yields the following: 1, 2, **3 ± a, 5 ± 2a, 8 ± 4a, 4 ± 7a, 3 ± 3a, 7 ± 2a, 1 ± 6a, 8 ± 9a, 9 ± 7a,**
**8 ± 8a, 8 ± 7a, 7 ± 7a, 6 ± 6a, 4 ± 5a, 1 ± 3a, 5 ± 9a, 6 ± 4a, 2 ± 5a, 8 ± a, 1 ± 7a, 9 ± 9a, 1 ± 8a,**
**1 ± 9a, 2 ± 9a;** 3 ± a, 5 ± 2a, 8 ± 4a ...

Again a 24 term repeating SDQ sequence independent of a.
Note that the coefficient of "a" for any term, is one less than the Fibonacci number in the previous term.

Each of the first 12 repeating terms when added to each of the remaining 12 terms equals **9 ± 7a**. Namely:
**3 ± a,  5 ± 2a, 8 ± 4a, 4 ± 7a, 3 ± 3a, 7 ± 2a, 1 ± 6a, 8 ± 9a, 9 ± 7a, 8 ± 8a, 8 ± 7a, 7 ± 7a,**
**6 ± 6a, 4 ± 5a, 1 ± 3a, 5 ± 9a, 6 ± 4a, 2 ± 5a, 8 ± a, 1 ± 7a, 9 ± 9a, 1 ± 8a, 1 ± 9a, 2 ± 9a**

## Fibo-like Plus 1

$F_{n+2} = F_{n+1} + F_n + 1$
1 2 4 7 3 2 6 9 7 8 7 7 6 5 3 9 4 5 1 7 9 8 9 9; 1 2 4
6 5 3 9 4 5 1 7 9 8 9 9

As with the Fibonacci series this is again a 24 term repeating series. . Notice that the each of the first 12 terms when added to each of the second 12 terms yields and **SDQ of 7**

## Fibo-like Minus 1

$F_{n+2} = F_{n+1} + F_n - 1$

1 2 2 3 4 6 9 5 4 8 2 9 1 9 9 8 7 5 2 6 7 3 9 2; 1 2 2 3
1 9 9 8 7 5 2 6 7 3 9 2

As with the Fibonacci series this is again a 24 term repeating series. Notice that the each of the first 12 terms when added to each of the second 12 terms yields and **SDQ of 2**.

# Chapter 6

## Pell Sequence

The Pell Sequence is given by $P_{n+2} = 2P_{n+1} + P_n$ where $P_1 = 1$, and $P_2 = 2$.

1, 2, 5, 12, 29, 70, 169, 408, 985, 2378, 5741, 13860, 33461, 80782, 195025, 470832, 1136689, 2744210, 6625109, 15994428, 38613965, 93222358, 225058681, 543339720; 1311738121, 3166815962, 7645370045, 18457556052, 44560482149

Taking the SDQ of each term yields:

**1, 2, 5, 3, 2, 7, 7, 3, 4, 2, 8, 9, 8, 7, 4, 6, 7, 2, 2, 6, 5, 7, 1, 9**; 1, 2, 5, 3, ...   A 24 term repeating sequence, where each of the second 12 terms when added to each of the first 12 terms has an SDQ = 9.

The 25th term divided by the 24th term yields 1,311,738,121/543,339,720 = 2.41421356237309504999992895789029 which is close to $1 + 2^{1/2}$
Around 2.414
Whose inverse is 0.41421356237309504859621290216357 which is close to $2^{1/2} - 1$.
Around 0.414
[Now $2^{1/2} = 1.41421356237309504880168872420969807856967187 5$]

In general: $P_{n+1}/P_n \gg 2^{1/2} + 1$ and $P_n/P_{n+1} \gg 2^{1/2} - 1$.
And of course: $(P_{n+1}/P_n)(P_n/P_{n+1}) = (1 + 2^{1/2})(2^{1/2} - 1) = 1$

## Pell-Like Sequence

We can start this sequence with 1, 1 instead of 1, 2.
The Pell Sequence is given by $P_{n+2} = 2P_{n+1} + P_n$ where $P_1 = 1$, and $P_2 = 1$.

The SDQ of this sequence is:

**1, 1, 3, 7, 8, 5, 9, 5, 1, 7, 6, 1, 8, 8, 6, 2, 1, 4, 9, 4, 8, 2, 3, 8**; 1, 1, 3, ...
Same 24 term repeating pattern and each of the first 12 added to each of the second 12 => 9.

**Tri-Pell Sequence**

The Tri-Pell Sequence is given by $P_{n+2} = 3P_{n+1} + P_n$; where $P_1 = 1$, and $P_2 = 2$.

1, 2, 7, 23, 76, 251, 829, 2,738, 9,043, 29,867, 98,644, 325,799, …
Taking the SDQ of each term yields:

**1, 2, 7, 5, 4, 8**; 1, 2, 7, 5, 4, 8; … A 6 term repeating sequence where each of the first 3 terms when added to each of the second 3 terms adds to an SDQ = 6.
The SDQ of the repeat block **1, 2, 7, 5, 4, 8** equals 27=>9.

The 12th Tri-Pell term when divided by the 11th term is 3.30277 and its inverse is 0.30277.
Now $3/2 + 13^{1/2}/2 \approx 3.30277$   and $13^{1/2}/2 - 3/2 \approx 0.30277$

In general: $P_{n+1}/P_n \approx 3/2 + 13^{1/2}/2$ and $P_n/P_{n+1} \approx 13^{1/2}/2 - 3/2$

Therefore: $(P_{n+1}/P_n)( P_n/P_{n+1}) = (3/2 + 13^{1/2}/2)( 13^{1/2}/2 - 3/2) = 1$

**Quadra-Pell Sequence**

The Quadra-Pell Sequence is given by $P_{n+2} = 4P_{n+1} + P_n$; where $P_1 = 1$, and $P_2 = 2$.

1, 2, 9, 38, 161, 682, 2,889, 12,238, 51,841, 219,602, 930,249, 3,940,598, 16,692,641…

The SDQ of this sequence is;
**1, 2, 9, 2, 8, 7, 9, 7**; 1, 2, 9, 2,
An 8 term repeating sequence where each of the first 4 terms when added to each of the last 4 terms add to an SDQ = 9.

The 12th Quadra-Pell term when divided by the 11th term is 4.236067977498 and its inverse is 0.236067977499.
Now $2 + 5^{1/2} \approx 4.236067977499$ and $5^{1/2} - 2 \approx 0.236067977499$.

In general: $P_{n+1}/P_n \approx 2 + 5^{1/2}$ and $P_n/P_{n+1} \approx 5^{1/2} - 2$
Therefore: $(P_{n+1}/P_n)( P_n/P_{n+1}) = (2 + 5^{1/2})( 5^{1/2} - 2) =$

Add two digit sequences

**Golden Proportion Patterns**

The Golden Proportion, phi $\phi$ = 1.610834 and its inverse $1/\phi$ = 0.618034. Notice that $1 + 1/\phi = \phi$, and that $1 + \phi = \phi^2$. In other words, we have the same decimal for both $\phi$ and its inverse. Thus, the pattern here is a number like phi that has an integer, here **1**, followed by a decimal such that the inverse of this number is exactly the same as the decimal part. In other words, we have integer, n plus decimal x, where the inverse is equal to x.

$m + x = 1/x$, where m = 1, 2, 3…

Multiplying by x and collecting all terms we get the quadratic: $x^2 + mx - 1 = 0$.

This has the solutions: $x = [-m \pm (m^2 + 4)^{1/2}]/2$

m = 1    $x = [-1 \pm (5)^{1/2}]/2 = \phi$, and the solutions are: 0.618… and its inverse **1.618**… Fibonacci-like

m = 2    $x = -1 \pm (2)^{1/2}$, and the solutions are: 0.414… and **2.414**… Pell

m = 3    $x = [-3 \pm (13)^{1/2}]/2$ and the solutions are: 0.30277… and **3.30277**…

Tri-Pell

m = 4    $x = -2 \pm (5)^{1/2}$ and the solutions are: 0.23606… and **4.23606**… Quadra-Pell

m = 5    $x = [-5 \pm (29)^{1/2}]/2$

m = 6    $x = -3 \pm (10)^{1/2}$

m = 7    $x = [-7 \pm (53)^{1/2}]/2$

m = 8    $x = -4 \pm 17^{1/2}$

m = 9    $x = [-9 \pm 85^{1/2}]/2$

m = 10   $x = -5 \pm 26^{1/2}$

This general result is for the term ratios in the sequences: $P_{n+2} = mP_{n+1} + P_n$ where $P_1 = 1$, and $P_2 = 2$.

**INVESTIGATE**

Starting with 5, every other Fibonacci number $\{0,1,1,2,3,\mathbf{5},8,\mathbf{13},21,\mathbf{34},55,\mathbf{89},...\}$ is the length of the hypotenuse of a right triangle with integral sides, or in other words, the largest number in a Pythagorean triple. The length of the longer leg of this triangle is equal to the sum of the three sides of the preceding triangle in this series of triangles, and the shorter leg is equal to the difference between the preceding bypassed Fibonacci number and the shorter leg of the preceding triangle.

The first triangle in this series has sides of length 5, 4, and 3. Skipping 8, the next triangle has sides of length 13, 12 $(5+4+3)$, and 5 $(8-3)$. Skipping 21, the next triangle has sides of length 34, 30 $(13+12+5)$, and 16 $(21-5)$. This series continues indefinitely and approaches a limiting triangle with edge ratios:
$5^{1/2} : 2 : 1$

**SDQ and $x^n + y^n = z^n$**

Look at $x^8 + y^8 = z^8$

look also at: $\mod(n)$ and $SDQ(\mod(n))$

# Chapter 7
## Golden Angles and the Great Pyramid

### *SURVEYING THE PAST*

To survey the past we need to extend the scope of modern surveying in both time and space. Through the employment of new techniques, the fields of astronomy and archeology are gaining results that could provide a treasure trove of scientific and financial discoveries. Underlying both fields there is a need for advanced professional surveying. In the past these fields have been managing well independent of one another but now it is this writers contention that all could benefit by a marriage of astronomy, archeology and surveying. It is the purpose of this column to attempt such a marriage by investigating areas of mutual interest in all three areas. Archeology spans the past, astronomy the past present and future, and extended surveying the tool to quantify all three.

To ancient Egyptians each annual flooding and receding of the Nile waters exposed a new landscape of nutrient rich soil providing high yield crops for a mighty civilization. How was the annual surveying accomplished in order to ensure an equitable distribution of farmland? How were these Bronze Age people able to survey their entire land to accuracy rivaling modern methods?

Indeed how was the Great Pyramid built with its center placed precisely bisecting the northern hemisphere and on 13 acres of bedrock to an elevation precision of a fraction of an inch? How did they use the sun to measure the land? How did they map the skies onto their land with the Milky Way as the Nile, and pyramids as the stars of Orion? Is it possible to use this information to locate buried ancient Egyptian treasure?

There is increasing evidence that ancient peoples such as the Egyptians and Sumerians utilized the sky to mark earthly locations. Not only did the ancient Egyptians know that the earth was a spheroid, they actually used longitude spacing at the equator to make corrections at their latitude spanning 7 degrees. To them the Nile represented the Milky Way and the three pyramids at Gizeh matched to spacing of the stars in the belt of Orion.

The Golden Proportion reveals one of the fundamental fabrics of the universe. Symbolized by the Greek letter phi, $\phi$ it represents the numerical pattern for growth and beauty. An irrational number, $\phi$ when added to the number 1, becomes its own square, $\phi^2$. A startling interpretation is that by adding 1 it increases its dimensionality. Namely, $1 + \phi = \phi^2$.

The Golden Proportion is alluded to in the New Testament, The Book of John, which opens as follows:
"In the beginning was the Word,
  and the Word was with God,
  and the Word was God,
  as in the beginning."

Taken as a mathematical "word" problem, letting W denote Word, G denote God, yields:
In the beginning was W, (W),
And W was with G, (W + G),
And W was G, (G),
As in the beginning, (W).

Viewed as a proportion this becomes: $W : (W + G) :: G : W$.
Cross multiplying gives: $W^2 = G(W + G)$. Taking G god to be 1 yields:
$W^2 = W + 1$. This quadratic equation has two solutions, $W = (1 \pm 5^{1/2})/2$.
The positive root solution is called $\phi$ the Golden Proportion.
$\phi = (1 + 5^{1/2})/2 \approx$
1.61803398874989484820458683436563811772030917980576286213544862270526046281890244970720720418939 1137+

Now $\phi^2 = [(1 + 5^{1/2})/2][(1 + 5^{1/2})/2] = (1/4)(6 + 2 \times 5^{1/2}) = 1 + \phi$, as expected.
The other root, written as $\phi' = (1 - 5^{1/2})/2 \approx$
-0.61803398874989484820458683436563811772030917980576286213544862270526046281890244970720720418939 1137+

We immediately note that $-\phi' = \phi - 1$.
Also that $1/\phi = \phi - 1 = -\phi'$

**Fibonacci and SDQ**

Clearly the Fibonacci Series as given is a non-repeating series. However, if we apply modulo 9 to the series, where each number is reduced to a single digit (which I call SDQ, or Single Digit Quality), and

find this SDQ series does have a repeating period of 24 terms.

SDQ Fibonacci: **1, 1, 2, 3, 5, 8, 4, 3, 7, 1, 8, 9, 8, 8, 7, 6, 4, 1, 5, 6, 2, 8, 1, 9**; 1, 1, 2, 3, 5,

This repeating series has five 1's and 8's, two 2's, 3's, 4's, 5's, 6's, 7's, and 9's. These have a total SDQ of 9. This is one of the secret patterns of nature (and stock market analyses).

**Twenty-four Term Repeating Pattern**

Actually we need only the first half (12 terms) to determine the full pattern. Notice that the 13th term and the first term add to 9, as do all the following terms, 14th and 2nd, etc. Thus we need only add to each of the first 12 terms the corresponding terms that add to an SDQ of 9.

**A Fibonacci named Moebius**

The twenty four SDQ repeating terms form the foundation pattern of nature. We can construct a Moebius strip containing these terms so that each of the first 12 terms has its corresponding term on the opposite to form the complete 24 term pattern. Each number then together with its opposite side number add to and SDQ of 9. We also find that writing out all of the terms on one side and then twisting the strip to form the Moebius strip, the terms on opposite sides add to and SDQ of 9 and are opposite down. Could this be the actual pattern of string theory that forms the fundamental nature of all things?

**The Golden Section**

This refers to the division of any line into two sections that are in the $\phi$ proportion. Consider line segment PGA. We can obtain the point G that creates this proportion by the following

P _____G_____ A

If PA = 1, then since $\phi^{-2} + \phi^{-1} = \phi^0 = 1$, then GA = $\phi^{-2} \approx 0.382$, and PG = $\phi^{-1} \approx$ 0.618. As a check the golden proportion states that AC:AB::AB:BC = $\phi$. Thus 1: $\phi^{-1}$ :: $\phi^{-1}$: $\phi^{-2}$ = $\phi$, as expected.

**The Circle Golden Angle**

The Golden Angle refers to the division of the circle into two parts that are in the φ relationship,

φ$^{-2}$: φ$^{-1}$. Therefore, x:(360º - x)::0.382:0.618 so that x = 137.52º and 360 − x = 222.48º. The smaller of the two is taken as the Golden Angle = 137.52º.

## Golden Angles

Taking our clue from the golden line section, (PA/PG = φ), we raise the vertical line, PGA, where A is at the top, point G locates the golden proportion and Point P is on the ground at a distance x from point F. We now draw lines FA and FG thus forming two right triangles, FAP and FGP. Finally let us label angle AFP = α, and angle GFP = β. We seek what we call the "Golden Angles," the relationship between α and β that divide the Golden Proportion, for all x values.

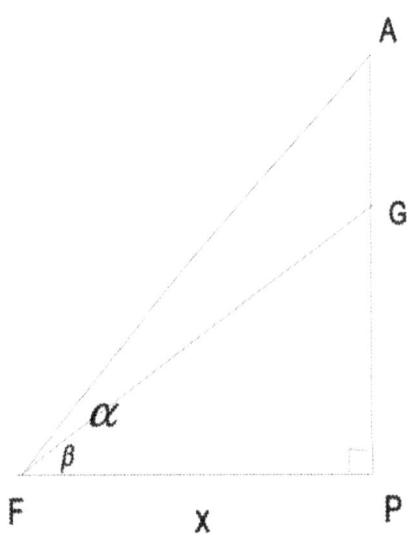

PA = x tan α, PG = x tan β, so that PA/PG = φ = x tanα /x tan α = tan α /tan β.
Thus, tan β = (tan α)/φ. Finally, we get the required relationship β = tan$^{-1}$ ((tan α)/φ). As seen from Table there is a set of golden angles such that their difference, α - β, is a maximum. This maximum occurs between the α values of 50 and 55 degrees, intentionally left blank in the following table. The significances of that unique angle will be examined in detail.

| α | β | α - β | α + β |
|---|---|---|---|
| 5 | 3.09 | 1.91 | 8.09 |
| 10 | 6.215 | 3.785 | 16.215 |
| 15 | 9.397 | 5.603 | 24.397 |

| 20 | 12.675 | 7.325 | 32.675 |
| 25 | 16.08 | 8.92 | 41.08 |
| 30 | 19.636 | 10.364 | 49.636 |
| 36 | 24.18 | 11.82 | 60.18 |
| 40 | 27.41 | 12.59 | 67.41 |
| 45 | 31.72 | 13.28 | 76.72 |
| 50 | 36.37 | 13.63 | 86.37 |
| | Intentionally Left Blank | | |
| 55 | 41.44 | 13.56 | 96.44 |
| 60 | 46.95 | 13.05 | 106.95 |
| 65 | 52.98 | 12.02 | 117.98 |
| 70 | 59.50 | 10.5 | 129.50 |
| 72 | 62.27 | 9.73 | 134.27 |
| 75 | 66.56 | 8.44 | 141.56 |
| 80 | 74.08 | 5.92 | 154.08 |
| 85 | 81.94 | 3.06 | 166.94 |

Between the angles of 50 and 55 the $\alpha - \beta$ has a maximum and $\alpha + \beta$ seems to be around 90. We can calculate the maximum by taking the derivative of the $\alpha - \beta$ with respect to $\alpha$, in other words to evaluate $(d/d\alpha)(\alpha - \beta) = 0$.

$(d/d\alpha)[\alpha - (\tan^{-1} (\phi^{-1}\tan \alpha)] = 0$. Performing this operation we get (calling $\alpha_{max}$ the value of $\alpha$ where $\alpha - \beta$ has its maximum, and $\beta(\alpha_{max})$ the corresponding value of $\beta$.

$\alpha_{max} = \mathbf{sin^{-1}} (\phi - 1)^{1/2} = \sin^{-1} 0.7861 = 51.82º$. For this special value $\beta$, $\beta(\alpha_{max}) = 38.17º$ so that

$\alpha_{max} + \beta(\alpha_{max}) = \mathbf{89.99º \approx 90º}$ **(the theoretical complimentary pyramid angles).**

## The Great Pyramid Angle

The Great Pyramid at Giza was designed so that the perimeter of its square base was equal to the circumference of a circle whose radius is the Pyramid' height. (This is called Squaring the Circle).

The following is a representation of the Great Pyramid at Giza. The base BECD is a square with side b. Point P is at the center of the base; AP is the pyramid height h, and the length of each pyramid side is s.

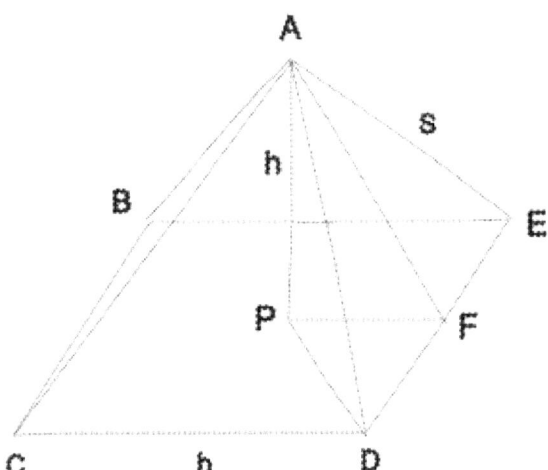

Draw point F that bisects base side DE.

Draw lines PD and PF. PF = DF = FE = b/2.

The perimeter of the square base is 4b. The circumference of the circle, whose radius is height h, is 2πh.

Thus, according to the design condition 4b = 2πh, so that PF = b/2 = πh/4.

**The Pyramid Angle AFP = tan$^{-1}$ h/(πh/4) = tan$^{-1}$ 4/π = 51.85°.**

## *Comparing The Great Pyramid and Golden Angles*

The Great Pyramid Angle AFP = tan$^{-1}$ 4/π

The Golden Angle AFP = sin$^{-1}$ (ϕ - 1)$^{1/2}$

Using tan$^{-1}$x = sin$^{-1}$(x/(1 + x$^2$)$^{1/2}$) we are now comparing (ϕ - 1)$^{1/2}$ to ((4/π)/(1 + (4/π)$^2$)$^{1/2}$. Squaring both expressions they are ϕ - 1 and ((4/π)$^2$/(1 + (4/π)$^2$)).

Therefore, ϕ = 1 + ((4/π)$^2$/(1 + (4/π)$^2$)) = 1 + 1.6211389/2.6211389 = 1.618486467190809…

Compare with the actual value for ϕ = 1.618033988749895…

They are within a difference of less than 0.2%

Now 6ϕ$^2$ = 15.70820393249937… and 5π = 15.70796326794897… a difference of less than 0.002%.  **Φ✿Φ ≈ ☆✝**

Expressing ϕ in terms of π, ϕ -1 becomes (5π - 12)/6, since 1 + ϕ = ϕ$^2$.

Leaving the final form of the comparison between 16/(π$^2$ + 16) to (5π - 12)/6.

16/(π$^2$ + 16) = 0.6184864581588363

(5π - 12)/6 = 0.6179938779914943 which differ by less than 0.085%.

We therefore conclude that the two angles are practically identical.

## Conclusions

It appears that the great pyramid angle was chosen to visually portray the golden proportion. The great golden pyramid angle and its complement are strikingly demonstrated by the great pyramid. The angle formed by the pyramid side and the vertical is this remarkable set of golden angles. It is not only the angle that is seen, but also the angle made with the unseen vertical that is so remarkable. It appears that this set of angles was no accident, but was analytically chosen to represent the most beautiful of all proportions at their maximum.

## Addendum

Note that every vertical line drawn within triangle APF, such as A'G'P', is a golden section. We now have a visual rendition of all such golden sections.

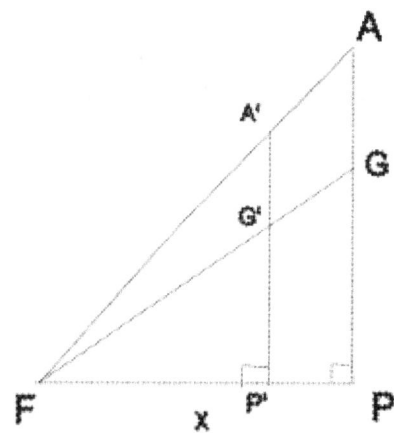

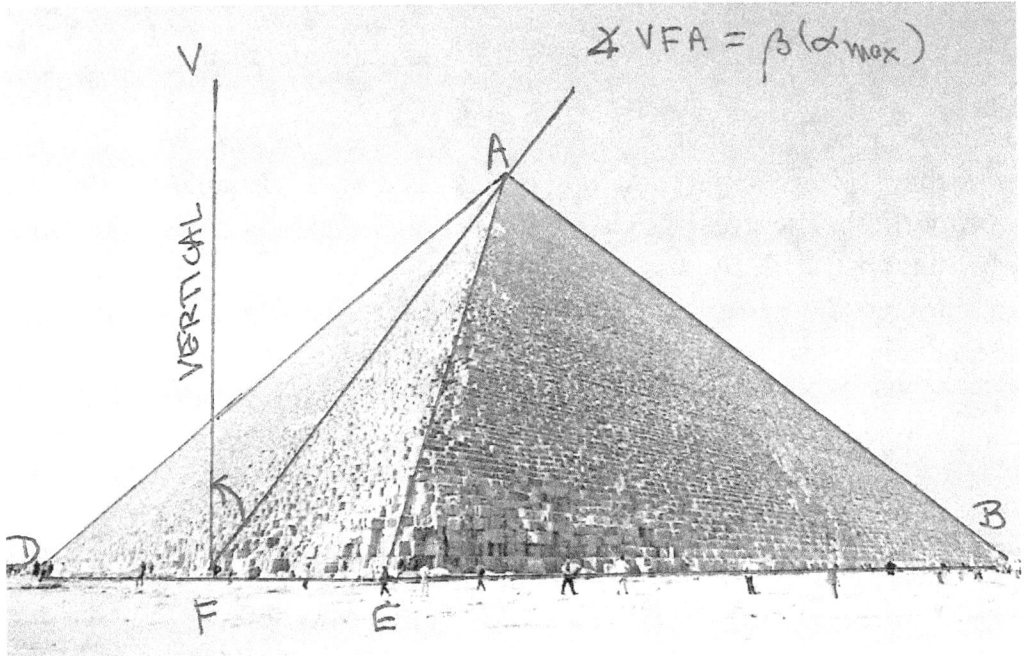

THE GREAT PYRAMID AT GIZA with a hand drawing illustrating the angle, **VFA, THE GOLDEN PYRAMID ANGLE COMPLEMENT**. - Pyramid photo by Nina Aldin Thune, 2005

**Pythagorean Golden Pyramid Angles**

The following right triangle APF where PG is perpendicular to AF illustrates the Pythagorean Golden Pyramid Angle **51.83°** and its golden complementary angle 38.17°. The golden proportion has the property that $1 + \phi = \phi^2$ and the Golden Fibonacci series is

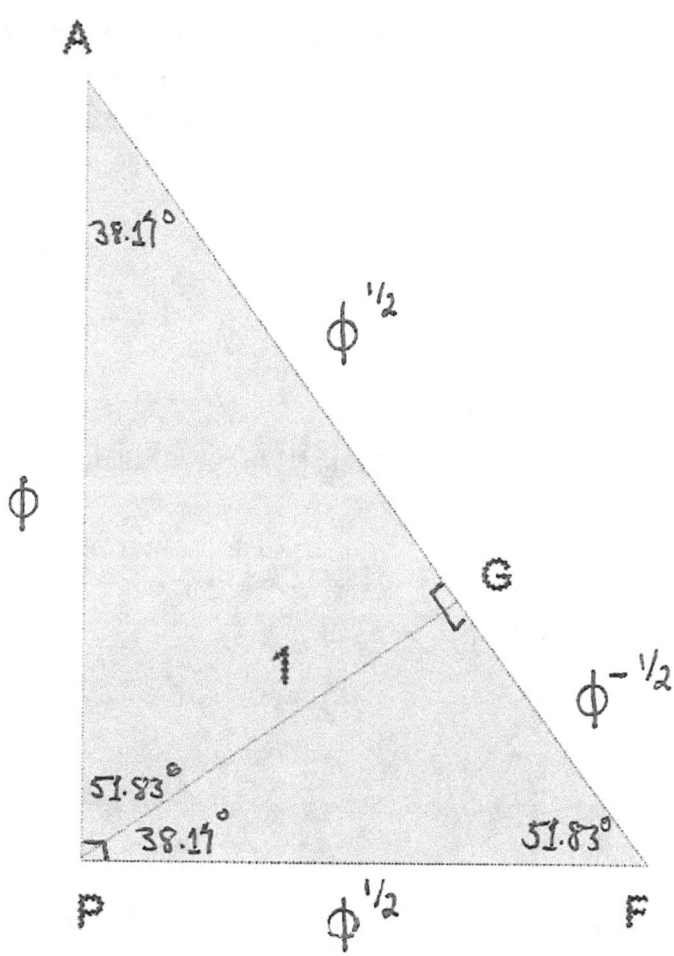

**The Golden Proportion, Fibonacci and SDQ**

In the Fibonacci series each term is the sum of the two preceding terms. Thus, 1, 1, 2, 3, 5, 8, 13, 21, 34, 55, … We notice that any term when divided by the immediately preceding term approaches the Golden Proportion, $\phi$, 55/34 ≈ 1.618. Any term divided by the following term approaches the inverse of the Golden Proportion 1/$\phi$, 34/55 ≈ 0.618.

This is clearly illustrated with the Golden Fibonacci Series: …, $\phi^{-7}$, $\phi^{-6}$, $\phi^{-5}$, $\phi^{-4}$, $\phi^{-3}$, $\phi^{-2}$, $\phi^{-1}$, $\phi^0$, $\phi^1$, $\phi^2$, $\phi^3$, $\phi^4$, $\phi^5$, $\phi^6$, $\phi^7$, …. This is a remarkable series of infinite increasing dimensionality, a kind of Hilbert Space.

In the Fibonacci series each term is the sum of the two preceding terms. Thus, 1, 1, 2, 3, 5, 8, 13, 21, 34, 55, … We notice that any term when divided by the immediately preceding term approaches the Golden Proportion $\phi$, 55/34 ≈ 1.618. Any term divided by the following term approaches the inverse of the Golden Proportion, $\phi^{-1}$ = 34/55 ≈ 0.618.

This is clearly illustrated with the Golden Fibonacci Series: …, $\phi^{-7}$, $\phi^{-6}$, $\phi^{-5}$, $\phi^{-4}$, $\phi^{-3}$, $\phi^{-2}$, $\phi^{-1}$, $\phi^0$, $\phi^1$, $\phi^2$, $\phi^3$, $\phi^4$, $\phi^5$, $\phi^6$, $\phi^7$, …. This is a remarkable series of infinite increasing dimensionality, a kind of Hilbert Space.

**The Golden Proportion, Fibonacci and SDQ**

Since 1 + $\phi$ = $\phi^2$ we get the following relationships: [Where => means taking the SDQ value]

| | | |
|---|---|---|
| $\phi^2 = 1 + \phi$ | $\phi^{14} => 8 + 8\phi$ | $\phi^{26} => 1 + \phi$ 24 term SDQ period |
| $\phi^3 = \phi^2 + \phi = 1 + 2\phi$ | $\phi^{15} => 8 + 7\phi$ | $\phi^{27} => 1 + 2\phi$ |
| $\phi^4 = \phi^3 + \phi^2 = 2 + 3\phi$ | $\phi^{16} => 7 + 6\phi$ | … |
| $\phi^5 = 3 + 5\phi$ | $\phi^{17} => 6 + 4\phi$ | |
| $\phi^6 = 5 + 8\phi$ | $\phi^{18} => 4 + \phi$ | |
| $\phi^7 = 8 + 13\phi => 8 + 4\phi$ | $\phi^{19} => 1 + 5\phi$ | |
| $\phi^8 => 4 + 3\phi$ | $\phi^{20} => 5 + 6\phi$ | |
| $\phi^9 => 3 + 7\phi$ | $\phi^{21} => 6 + 2\phi$ | |
| $\phi^{10} => 7 + \phi$ | $\phi^{22} => 2 + 8\phi$ | |
| $\phi^{11} => 1 + 8\phi$ | $\phi^{23} => 8 + \phi$ | |
| $\phi^{12} => 8 + 9\phi$ | $\phi^{24} => 1 + 9\phi$ | |
| $\phi^{13} => 9 + 8\phi$ | $\phi^{25} => 9 + \phi$ | |

We immediately see that $\phi^{26} = \phi^2$, $\phi^{27} = \phi^3$, and so forth.

It will also be noticed that we need only the first half (12 terms, $\phi^2$ to $\phi^{13}$) of the repeating SDQ series. Each of the next terms (red) add to the corresponding term (blue) to become $9 + 9\phi$. Thus, $\phi^2 = 1 + \phi$ when added to $\phi^{14} => 8 + 8\phi$ becomes $9 + 9\phi$.

Also note that $SDQ(\sum \phi^n$ (from n=2 to n = 13)) = $7 + 5\phi$, and $SDQ(\sum \phi^n$ (from n=14 to n = 25)) = $2 + 4\phi$. So that $SDQ(\sum \phi^n$ (from n=2 to n = 25) = $9 + 9\phi = 9\phi^2$ the SDQ sum of the entire Fibonacci Series.

Finally, since any number when multiplied by 9 has an SDQ of 9, the final SDQ of the Fibonacci Golden Series is 9.

[We can perform a similar analysis to the negative exponentials.]

## Mystical Math

We now decrypt the symbolism and meaning behind the following relationship

$$\star\; \text{♀} \approx \Phi\; \text{✡}\; \Phi$$

Each symbol represents a number

The symbol on the left $\star$ means 5

♀ The second symbol, the Ankh, here show the circumference (of a circle), divided ♀by the diameter which is equal to $\pi$

$\Phi$ is the third symbol, phi which equals $(1 + 5^{1/2})/2$

Thus $5\,\pi \approx 6\,\Phi^2$

$5 \times 3.14159$ is approximated by $6 \times 1.618^2$

15.708 is approximated by 15.708

Pretty good approximation.

Another good approximation

113355 when split in two and the second half is divided by the first half becomes 355/113 which equals 3.141593 the same as pi for this number of places. A pretty good approximation.

# Chapter 8

## Perrin Sequence

$P(n) = P(n-2) + P(n-3)$, where $P(0) = 3$, $P(1) = 0$, and $P(2) = 2$.

3, 0; **2, 3, 2, 5, 5, 7,** 10, 12, **17,** 22, **29,** 39, 51, 68, 90, **119,** 158, **209, 277, 367,** 486, 644,

853, 1130, **1497,** 1983, 2627, 3480, 4610, 6107, 8090, 10717, 14197, 18807, 24914,

33004, 43721, 57918, 76725; 101639, 134643, 178364,

**Perrin Numbers in bold italic are prime numbers.**

101639/76725 = 1.32471814923427826653633310524601...

178364/134643 = **1.324717957858930653654478881895...** = w.

This satisfies $1 + w = w^3$. Similar to $1 + \phi = \phi^2$.

w ≈ 1.324718

76725/101639 = 0.754877556843337695176064306024 26

134643/178364 = **0.754877665896705613240339978919**51

This is a 39 term repeating sequence in SDQ. P(2) ... P(40). P(41) ... P(79).

3, 0; **2, 3, 2, 5, 5, 7,** 1, 3, 8, 4, 2, 3, 6, 5, 9, 2, 5, 2, 7, 7, 9, 5, 7, 5, 3, 3, 8, 6, 2, 5, 8, 7, 4,

6, 2, 1, 8, 3, 9; **2, 3, 2, 5, 5, 7,** ...

All the terms including the center (20 terms) add to SDQ = 9. As do all of the remaining terms, so that the entire repeat block has a SDQ of 9.

| Number | Frequency |
|--------|-----------|
| 1 | 2 |
| 2 | 7 |
| 3 | 6 |
| 4 | 2 |
| 5 | 7 |
| 6 | 3 |
| 7 | 5 |
| 8 | 4 |
| 9 | 3 |

Total number times frequency = 189 => 9.

**Perrin Primes**

A Perrin prime is a Perrin number that is prime. The first few Perrin primes are (sequence A074788 in OEIS):

2, 3, 5, 7, 17, 29, 277, 367, 853, 14197, 43721, 1442968193, 792606555396977, 187278659180417234321, 66241160488780141071579864797,

Perrin's sequence.
0;
2,3,2,5,5,7,10,12,17,22,29,39,51,68,90,119,158,209,277,367,486,644,853,1130,1497,
1983, 2627,3480,4610,6107,8090,10717,14197,18807,24914,33004,43721,57918,
76725;
101639,134643, 178364,236282,313007,414646,549289,727653,963935,
1276942, 1691588,2240877,2968530,3932465,
5209407,6900995,9141872,12110402,16042867,21252274,28153269,37295141,494055
43,65448410,
86700684,114853953,152149094,201554637,267003047,353703731,468557684,62070
6778,822261415,
1089264462,1442968193,1911525877,2532232655,3354494070,4443758532,58867267
25,7798252602,
10330485257,13684979327,18128737859,24015464584,31813717186,42144202443,55
829181770,
73957919629,97973384213,129787101399,171931303842,227760485612,3017184052
41,399691789454,
529478890853,701410194695,929170680307,1230889085548,1630580875002,...

# Chapter 9

# Prime Numbers

## INTRODUCTION

In this chapter we examine properties of prime numbers in terms of their SDQ Single Digit Qualities, a technique that is based upon ancient Hebrew gematria. Prime numbers play prominent roles in music theory, as regular numbers, and as markers on the fundamental octave. Their function within the tonal index has far reaching applications in both creating and in setting limits on growth. For the first time, a prime number generator is given to generate all of the prime numbers greater than 3. Chapters 10-12 will detail prime number tonal music characteristics.

Prime numbers are particularly related to the perfect number 6, and to the 6 Elohim of Genesis. These aspects will be analyzed in chapter 22. We begin with 12 definitions, where the symbol ≡ means "is defined as."

1- Integer ≡ a positive or negative number including zero: -5, -4, -3, -2, -1, 0, 1, 2, ...

2- Non negative integer ≡ 0, 1, 2, 3, 4, 5, 6, ... .

3- Positive integer ≡ 1, 2, 3, 4, 5, 6, ... .

4- Rational number ≡ one that can be formed by division of two integers, a/b, where b is not equal to 0, b ≠ 0.

5- Prime number, p ≡ any integer greater than 1, that is not divisible by any other number except itself and 1. Thus, p = 2, 3, 5, 7, 11, ... .

6- Prime prime number ≡ a prime prime number is a prime number that also has an SDQ that is prime.

7- Composite number ≡ a number greater than 1 and not prime. Thus, 4, 6, 8, 9, 10, 12, ... .

8- In composite numbers ≡ all the prime numbers including 0 and 1. Thus, 0, 1, 2, 3, 5, 7, 11, ... .

9- In composite non negative numbers ≡ all in composite numbers excluding 0. Thus, 1, 2, 3, 5, 7, 11, ... .

10- Congruent modulo d ≡ any two integers are congruent modulo d, if they leave the same remainder when divided by d. The following statements are all equivalent:

        1. a is congruent to b modulo d

        2. a = b (mod d)

        3. a = b + nd, for some integer n

        4. d divides a - b.

11- Regular numbers ≡ numbers that can be formed from $2^p3^q5^r$. The base numbers are all prime numbers, 2, 3, 5, ... . The exponents are non negative integers, 0, 1, 2, 3, 4, 5, ... .

12- Tonal Index ≡ the largest value of $2^p3^q5^r$ in a particular <u>musical</u> system. The tonal index limits the total number of allowable tones.

## PRIME NUMBER THEOREMS

Now let us consider some theorems about prime numbers.

**THEOREM: Every positive integer, except 1, is a product of primes.**

This theorem has the consequence that either a number is prime, and therefore stands alone, or it is composite, and is composed of products of prime numbers. Thus, the number 4, is not a prime and is composed of $2 \times 2 = 2^2$. The number $27 = 3 \times 3 \times 3 = 3^3$. The number $128 = 2 \times 2 \times 2 \times 2 \times 2 \times 2 \times 2 = 2^7$. Therefore, either numbers are prime and is a stand alone number, or it is of the form $2^p3^q5^r7^s...$ . Where each base is a prime number.

**Observations**:

There are 10 digits: 0 ,1, 2, 3, 4, 5, 6, 7, 8, and 9.

There are 6 in composite digits are: 0, 1, 2, 3, 5, 7, and they add to 18 ==> 9. Thus, the sum of the in composite digits has an SDQ of 9.

There are 4 composite digits: 4, 6, 8 and 9, and they add to 27==>9. Thus, the sum of the composite digits has an SDQ of 9. The product of the 4 incomposite digits is 1728==>18==>9, and also with an SDQ of 9.

There are 4 prime digits: 2, 3, 5, and 7, and they add to 17==>8. The product of the 4 prime digits is 210==>3.

## FERMAT'S THEOREM-The Fundamental Theorem of Arithmetic

**Any integer can be resolved, in one way only, into a product of primes.** (N.B. Had we included 1 as a prime then this theorem would not be true). Notice that we are not using incomposite numbers which include the primes plus 0 and 1. Thus, $666 = 2 \times 3 \times 3 \times 37 = 2^13^237^1$.

**THEOREM: (The Prime Number Theorem). The number of primes not exceeding x is asymptotic to x/ln x.** If $\pi(x)$ denotes the number of primes not exceeding x, and ln means the natural log of x, then

$$\pi(x) \sim x/\ln x.$$

N.B. Asymptotic to here means: approaches x/ln x as x becomes very large.

**THE PRIME NUMBERS**

 The prime numbers, incomposite numbers excluding 0 and 1, are: 2, 3, 5, 7, 11, 13, 17, 19, 23, 29, 31, 37, 41, 43, 47, 53, 59, 61, 67, 71, 73, 79, 83, 89, 97, 101, 103, 107, 109, 113, 127, 131, 137, 139, 149, 151, 157, 163, 167, 173, 179, 181, 191, 193, 197, 199, 211, 223, 227, 229, 233, 239, 241, 251, 257, 263, 269, 271, 277, 281, 283, 293, 307, 311, 313, 317, 331, 337, 347, 349, 353, 359, 367, 373, 379, 383, 389, 397, 401, 409, 419, 421, 431, 433, 439, 443, 449, 457, 461, 463, 467, 479, 487, 491, 499, 503, 509, 521, 523, 541, 547, 563, 569, 571, 577, 587, 593, 599, 601, 607, 613, 617, 619, 631, 641, 643, 647, 653, 659, 661, 673, 677, 683, 691, 701, 709, 719, 727, 733, 739, 743, 751, 757, 761, 769, 773, 787, 797, 811, 821, 823, 827, 829, 839, 853, 857, 859, 863, 877, 881, 883, 887, 907, 911, 919, 929, 937, 941, 947, 953, 967, 971, 977, 983, 991, 997, 1009, 1013, 1019, 1021, 1031, 1033, 1039, 1049, 1051, 1061, 1063, 1069, 1087, ... .

## PRIME NUMBER PROPERTY

 Every prime number $p \geq 5$, when divided by the perfect number 6, has a remainder of either 1 or 5. Mathematically, we can write $p \pm 1$ (mod 6). This does not mean that every number, n, that can be divided by 6 and has a remainder of 1 or 5, is a prime number. It does mean, however, that if the number is prime then it must have a remainder of 1 or 5 when divided by 6.

**Primary Rule: $p \pm 1$ (mod 6). Every prime number $p \geq 5$, when divided by 6, has a remainder of 1 or 5.**

 **As a consequence of this primary rule, $(6n \pm 1)$ will generate every prime number. It will also generate composite numbers, but will include every prime number.**

## PRIME NUMBER GENERATOR

 **To generate all the primes $\geq 5$: let $p_n$ be the primes $\geq 5$ in increasing order, $p_1 = 5$, $p_2 = 7$, $p_3 = 11$, $p_4 = 13$, ... . Let $q_i$ be the value of the $i^{th}$ term of the 2n terms generated by $(6n \pm 1)$. Let $N = \max(6n+1)$.**

 **1- write down the $(6n \pm 1)$ terms in increasing order, where n = 1, 2,**

 **3, ...**
 **2- cancel out all 5(5, 7, 11, 13, 17, 19, 23, 25, 29, 31, 35, ..., $q_j$) terms, where $5q_j \leq N$**
 **3- cancel out all 7(7, 11, 13, ..., $q_k$) terms that remain, where $7q_k \leq N$**

 **4- continue to cancel out all $p_n(p_n, q_s, q_{s+1}, ..., q_l)$ terms,**

where $p_n q_l \leq N$

5- the final cancellation will be $p_m(p_m)$, where $p_m p_m \leq N$.

6- the remaining terms are all the primes $\geq 5$.

**EXAMPLE: Calculate all the prime numbers for the first 100 numbers equal to or greater than 5.**

Here n = 50, N = 301.

Let us consider class 1, defined as (6n-1), where n = 1, 2, ..., 50. Class 1 values are: 5, 11, 17, 23, 29, 35, 41, 47, 53, 59, 65, 71, 77, 83, 89, 95, 101, 107, 113, 119, 125, 131, 137, 143, 149, 155, 161, 167, 173, 179, 185, 191, 197, 203, 209, 215, 221, 227, 233, 239, 245, 251, 257, 263, 269, 275, 281, 287, 293, 299. Notice that SDQ (6n-1) = 5, 2, 8.

Now let us consider class 2, defined as (6n + 1), where n = 1, 2, ..., 50. Class 2 values are: 7, 13, 19, 25, 31, 37, 43, 49, 55, 61, 67, 73, 79, 85, 91, 97, 103, 109, 115, 121, 127, 133, 139, 145, 151, 157, 163, 169, 175, 181, 187, 193, 199, 205, 211, 217, 223, 229, 235, 241, 247, 253, 259, 265, 271, 277, 283, 289, 295, 301. Notice that SDQ (6n+1) = 7, 4, 1.

Combining both classes arranged in numerical order yields: $t_i$ = (5, 7, 11, 13, 17, 19, 23, 25, 29, 31, 35, 37, 41, 43, 47, 49, 53, 55, 59, 61, 65, 67, 71, 73, 77, 79, 83, 85, 89, 91, 95, 97, 101, 103, 107, 109, 113, 115, 119, 121, 125, 127, 131, 133, 137, 139, 143, 145, 149, 151, 155, 157, 161, 163, 167, 169, 173, 175, 179, 181, 185, 187, 191, 193, 197, 199, 203, 205, 209, 211, 215, 217, 221, 223, 227, 229, 233, 235, 239, 241, 245, 247, 251, 253, 257, 259, 263, 265, 269, 271, 275, 277, 281, 283, 287, 289, 293, 295, 299, 301), where i = 1, ..., 100.

Together, both classes yield all the prime numbers, $p_n$ equal to and greater than $(6 \cdot 1 - 1) = 5$, and equal to and less than $(6 \cdot 50 + 1) = 301$. These 100 numbers also include composite numbers. The SDQ of these 100 numbers are the six sequentially repeating digits: 5, 7, 2, 4, 8, 1.

Class 1 contains 32 prime numbers, while class 2 contains 28. Thus, out of the 100 numbers of both classes, 60 are prime numbers. The 40 composite numbers can be struck out using the following method, noticing that the first 7 numbers in the combined classes are all primes.

Strike out the composite numbers 5(5, 7, 11, 13, 17, 19, 23, 25, 29, 31, 35, 37, 41, 43, 47, 49, 53, 55, 59). These must be composite numbers as they are formed from a product of two numbers. Notice that the composite numbers that we have just struck out are 5 times each of the first 19 numbers in the combined classes, limited by the product being equal to or less than 301. Since 5 x 59 = 295, and 5 x 61 = 305, the last number must be 59. This removes 19 of the 40 composite numbers and yields: 5, 7, 11, 13, 17, 19, 23, 29, 31, 37, 41, 43, 47, 49, 53, 59, 61, 67, 71, 73, 77, 79, 83, 89, 91, 97, 101, 103, 107, 109, 113, 119, 121, 127, 131, 133, 137, 139, 143, 149, 151, 157, 161, 163, 167, 169, 173, 179, 181, 187, 191, 193, 197, 199, 203, 209, 211, 217, 221, 223, 227, 229, 233, 239, 241, 247, 251, 253, 257, 259, 263, 269, 271, 277, 281, 283, 287, 289, 293, 299, 301. We have therefore struck out the numbers 5(5, 7, 11, 13, 17, 19, 23, 25, 29, 31, 35, 37, 41, 43, 47, 49, 53, 55, 59). Essentially we have struck out all numbers that are greater than 5 that end in a 5. This is done since any number greater than 5 that ends in a 5, is divisible by 5, and thus composite. This list includes 40 - 19 = 21 composite numbers, the first composite number being 7 x 7.

Now strike out all numbers of the form 7(7, 11, 13, 17, 19, 23, 29, 31, 37, 41, 43), where the numbers in parentheses are all successive numbers that remain after the first strike by 5, and that are equal to and greater than the prime number 7. The last number is 43 as it produces the largest product that is equal to or less than 301. We are therefore striking out 7 times the first 11 numbers in the remaining list that are equal to or greater than 7. In other words, all numbers divisible by the prime number 7, e.g. 49, 77, 91, 119, 133, 161, 203, 217, 259, 287, 301. This removes 11 of the remaining 21 composite numbers, and yields: 5, 7, 11, 13, 17, 19, 23, 29, 31, 37, 41, 43, 47, 53, 59, 61, 67, 71, 73, 79, 83, 89, 97, 101, 103, 107, 109, 113, 121, 127, 131, 137, 139, 143, 149, 151, 157, 163, 167, 169, 173, 179, 181, 187, 191, 193, 197, 199, 209, 211, 221, 223, 227, 229, 233, 239, 241, 247, 251, 253, 257, 263, 269, 271, 277, 281, 283, 289, 293, 299. This list now includes 21 - 11 = 10 composite numbers, the first composite number being 11 x 11.

Now strike out all numbers of the form 11(11, 13, 17, 19, 23), where the number in parentheses are all successive primes equal to and greater than the prime number 11, and the largest product is equal to or less than 301. In other words, we are striking out 11 times the first 5 numbers in the remaining list that are equal to or greater than 11. In other words, all numbers divisible by the prime number 11, e.g. 121, 143, 187, 209, 253. This removes 5 of the remaining 10 composite numbers, and yields: 5, 7, 11, 13, 17, 19, 23, 29, 31, 37, 41, 43, 47, 53, 59, 61, 67, 71, 73, 79, 83, 89, 97, 101, 103, 107, 109, 113, 127, 131, 137, 139, 149, 151, 157, 163, 167, 169, 173, 179, 181, 191, 193, 197, 199, 211, 221, 223, 227, 229, 233, 239, 241, 247, 251, 257, 263, 269, 271, 277, 281, 283, 289, 293, 299. This list now contains 10 - 5 = 5 composite numbers, the first composite number being 13 x 13.

Following this pattern, we next strike out all numbers of the form 13(13, 17, 19, 23), where again the numbers in parentheses are all successive primes equal to or greater than the prime number 13, and the largest product is equal to or less than 301. In other words, all numbers divisible by the prime number 13, e.g. 169, 221, 247, 299. This removes 4 of the remaining 5 composite numbers, and yields: 5, 7, 11, 13, 17, 19, 23, 29, 31, 37, 41, 43, 47, 53, 59, 61, 67, 71, 73, 79, 83, 89, 97, 101, 103, 107, 109, 113, 127, 131, 137, 139, 149, 151, 157, 163, 167, 173, 179, 181, 191, 193, 197, 199, 211, 223, 227, 229, 233, 239, 241, 251, 257, 263, 269, 271, 277, 281, 283, 289, 293. This list now contains 5 - 4 = 1 remaining composite number, being 17 x 17.

Finally, the last composite number is removed by striking out the next prime number 17, multiplied by itself, 17(17) = 289. Notice that we must stop here in the strikeouts since the next strikeout product 19(19) is larger than 301. This list now contains no composite numbers.

The final result is: 5, 7, 11, 13, 17, 19, 23, 29, 31, 37, 41, 43, 47, 53, 59, 61, 67, 71, 73, 79, 83, 89, 97, 101, 103, 107, 109, 113, 127, 131, 137, 139, 149, 151, 157, 163, 167, 173, 179, 181, 191, 193, 197, 199, 211, 223, 227, 229, 233, 239, 241, 251, 257, 263, 269, 271, 277, 281, 283, 293. These are the first 60 prime numbers that are equal to or less than 301.

## PRIME SQUARE MARKERS

We call the Prime Square Markers, PSM, the $p_n$ x $p_n$ values: 5x5, 7x7, 11x11, 13x13, ... . These numbers mark the start of the next strike out series. All of these prime square markers, with the exception of 5x5, are contained in class 2, (6n+1).

## THE NUMBER OF PRIMES NOT EXCEEDING 301.

According to the Prime Number Theorem, the number of primes $\pi(x)$ not exceeding x is
$\pi(x) \sim x/\ln x$. Thus, the number of primes not exceeding 301 is given by $\pi(301) \sim$ 301/ln 301. Now 301/ln 301 = 53. The actual number of primes is 60, so that the standard approximation differs by 13.2%.

## EUCLID'S SECOND THEOREM

According to Euclid=s Second Theorem: The number of primes is infinite.
Since all primes are contained within class 1 and class 2, namely, 6n-1 and 6n+1, it is clear that since n is infinite so are the number of primes generated by each class.

## SDQ VALUES OF PRIME NUMBERS

If we take the SDQ of all the prime numbers (incomposite numbers without the 0 and 1), we get the following: 2, 3, 5, 7, 2, 4, 8, 1, 5, 2, 4, 1, 5, 7, 2, 8, 5, 7, 4, 8, 1, 7, 2, 8, 7, 2, 4, 8, 1, 5, 1, 5, 2, 4, 5, 7, 4, 1, 5, 2, 8, 1, 2, 4, 8, 1, 4, 7, ... . Notice that other than the first 3, there is never a 3, 6 or 9.

**Rule 1: No prime number or incomposite number greater than 3, can have SDQ value of 3, 6, or 9.**

## SDQ OF (6n ± 1)

The values of 6n are: 6, 12, 18, 24, 30, 36, 42, 48, 54, ... . The SDQ of 6n are: 6, 3, 9, 6, 3, 9, 6, 3, 9, ... . Therefore, the SDQ values of (6n -1) are: 5, 2, 8, 5, 2, 8, 5, 2,, 8, ... . The SDQ values of (6n + 1) are: 7, 4, 1, 7, 4, 1, 7, 4, 1, ... . Notice that none of the SDQ of (6n ± 1) include 3, 6, or 9, as required by rule 1, and also that the SDQ values of both (6n - 1) and (6n + 1) include all the other (six) values, namely, 1, 2, 4, 5, 7, 8.

## THE DOUBLY PRIME NUMBERS

The numbers in italics have SDQ values that are also prime, namely: 11, 23, 29, 41, 43, 47, 61, 83, 101, 113, 131, 137, 151, 173, 191, 223, 227, 241, 263, 281, 311, 313, 317, 331, 353, 401, 421, 443, 461, 599, 601, 641, 797, 821, 887, 911, 977, 1013, 1019, 1031, 1033, 1051, ... For example, 11 => 2, and both 11 and 2 are prime. Another example, 23 => 5 and both 23 and 5 are prime. As a last example, 29 => 11 => 2, and all three, 29, 11 and 2 are prime numbers.

## SDQ VALUES OF THE DOUBLY PRIME NUMBERS

The SDQ values of the prime prime numbers are: 2, 5, 2, 5, 7, 2, 7, 2, 2, 5, 5, 2, 7, 2, 2, 7, 2, 7, 2, 2, 5, 7, 2, 7, 2, 5, 7, 2, 2, 5, 7, 2, 5, 2, 5, 2, 5, 5, 2, 5, 7, 7, ... .

**Rule 2: The SDQ values of the prime prime numbers are 2, 5, or 7.**

**Since the prime primes come from the class of all primes, and since the SDQ of (6n -1) are 5, 2, 8 and the SDQ of (6n + 1) are 7, 4, 1, there are more of class (6n-1), then of class (6n+1), in the prime primes.**

## PRODUCT OF CONSECUTIVE PRIMES

Consider the consecutive primes $p_1 = 2$, $p_2 = 3$, $p_3 = 5$, $p_4 = 7$, ..., $p_n$, ... . The product of consecutive prime numbers can be obtained by multiplying any prime number by all of its prior prime numbers. This can be written mathematically as $\Pi\ p_n = p_1 p_2 p_3 p_4...$ . These are: 2, 2x3, 2x3x5, 2x3x5x7, 2x3x5x7x11, 2x3x5x7x11x13, 2x3x5x7x11x13x17, 2x3x5x7x11x13x17x19, 2x3x5x7x11x13x17x19x23, 2x3x5x7x11x13x17x19x23x29, ... . These values are: 2,

6, 30, 210, 2310, 30030, 510510, 9699690, 223092870, 6469693230, 200560490130, 7420738134810, ... .

## SDQ OF PRODUCTS OF CONSECUTIVE PRIMES

The SDQ of these values, or SDQ[$\Pi$ $p_n$] are: 2, 6, 3, 3, 6, 6, 3, 3, 6, 3, 3, 3, 6, 6, 3, 6, 3, 3, 3, 6, 6, 6, 3, 6, 6, 3, 3, 6, 6, 3, 3, 6, 2, 3, 6, 6, 6, 6, 3, 6, 3, 3, 6, 6, 3, 3, 3, 3, ... .

**Rule 3: The SDQ of the product of consecutive primes excluding the first 2, are either 3 or 6.**

Comment: notice that the numbers created by the product of consecutive primes are not prime numbers, excluding the first 2. That this is so can be seen immediately from rules 1 and 3, since by rule 3, the SDQ values of the product of consecutive primes are either 3 or 6, and by rule 1, no prime can have an SDQ of 3, 6, or 9, so that the product of consecutive primes must be composite.

## GENERATING PRIMES FROM THE PRODUCT OF CONSECUTIVE PRIMES

The product of consecutive primes can be written mathematically as $\Pi$ $p_n$ = $p_1 p_2 p_3 p_4$... . If we add 1 to each of these products of consecutive primes, the new number cannot have any of the product primes as a factor. The new number, ($\Pi$ $p_n$ + 1), then is a prime number. Remember that $\Pi$ $p_n$ = $p_1 p_2 p_3 p_4$... .

## SDQ OF PRIME NUMBERS FORMED FROM THE PRODUCT OF CONSECUTIVE PRIMES

Since the SDQ of a sum is the sum of the SDQ, we can take the SDQ of the primes formed from the product of consecutive primes, SDQ($\Pi$ $p_n$ + 1) = SDQ($\Pi$ $p_n$) + SDQ(1). By rule 3, the SDQ of the product of consecutive primes, excluding the first 2, is either 3 or 6. Also SDQ($\Pi$ $p_n$ + 1) = SDQ($\Pi$ $p_n$) + SDQ(1), so that we get:

$$SDQ(\Pi\ p_n + 1) = 2 \times 3 + 1 \Longrightarrow 6 + 1 = 7$$
$$2 \times 3 \times 5 + 1 \Longrightarrow 3 + 1 = 4$$
$$2 \times 3 \times 5 \times 7 + 1 \Longrightarrow 3 + 1 = 4$$
$$2 \times 3 \times 5 \times 7 \times 11 + 1 \Longrightarrow 6 + 1 = 7$$
$$2 \times 3 \times 5 \times 7 \times 11 \times 13 + 1 \Longrightarrow 6 + 1 = 7$$
$$2 \times 3 \times 5 \times 7 \times 11 \times 13 \times 17 + 1 \Longrightarrow 3 + 1 = 4$$
$$2 \times 3 \times 5 \times 7 \times 11 \times 13 \times 17 \times 19 + 1 \Longrightarrow 3 + 1 = 4$$
$$\times 23 + 1 \Longrightarrow 6 + 1 = 7$$
$$\times 29 + 1 \Longrightarrow 3 + 1 = 4 ...$$

**Rule 4: The SDQ of the prime numbers formed from the product of consecutive primes plus 1, is either 4 or 7.**

## WILSON'S THEOREM

A number p is a prime number if $((p-1)! + 1)/p = I$, an integer. This integer, I, is either a prime or a product of primes.

If $p = 2$, then we get $((2-1)! + 1)/2 = 1$, and therefore 2 is a prime number.

If $p = 3$, then we get $((3-1)! + 1)/3 = 1$, and 3 is also a prime number.

If $p = 4$, then we get $((4-1)! + 1)/4 = 7/4$ not an integer, and 4 is not a prime number.

If $p = 5$, then we get $((5-1)! + 1)/5 = 1$.

If $p = 6$, then we get $((6-1)! + 1)/6 = 121/6$ not an integer.

If $p = 7$, then we get $((7-1)! + 1)/7 = 103 ==> 4$.

If $p = 8$, then we get $((8-1)! + 1)/8 = 5041/8$ not an integer.

If $p = 9$, then we get $((9-1)! + 1)/9 = 40321/9$ not an integer.

If $p = 10$ then we get $((10-1)! + 1)/10 = 362881/10$

If $p = 11$ then we get $((11-1)! + 1)/11 = 3628801/11 = 329891 ==> 32 ==> 5$.

If $p = 12$ then we get $((12-1)! + 1)/12 = 39916801/12 = $ non integer

If $p = 13$ then we get $((13-1)! + 1)/12! = 36846277 ==> 7$.

**Rule 5: SDQ(n!) = 9, where n ≥ 6.** [This is a special case that follows from Theorem 80, See Hardy and Wright, that $(j-2)! = a(\bmod j)$, $(j≥3)$, where a=1 or 0, according as j is prime or composite].

We can write Wilson's expression in the form $(p-1)! + 1 = pI$. We can take the SDQ of both sides, and since the SDQ of a sum is the sum of the SDQ's, and the SDQ of a product is the product of the SDQ's, we get $SDQ(p-1)! + SDQ(1) = SDQ(p) \times SDQ(I)$. By Rule 5, for $p ≥ 7$, the SDQ of $(p-1)!$ is 9. Thus, the left hand side of the equation, for $p ≥ 7$, becomes $9 + 1 = 10 ==> 1$. This result means that the right hand side must equal 1, or that $SDQ(p) \times SDQ(I) = 1$. By Rule 1, a prime number cannot have an SDQ of 3, 6, or 9. Thus, if p is a prime number, greater than 7, its SDQ must have a value of 1, 2, 4, 5, 7, or 8. Inserting these possible values into the right hand side, we get the condition that $(1, 2, 4, 5, 7, 8) \times SDQ(I) = 1$. This condition means that for the possible SDQ values of p of 1, 2, 4, 5, 7, or 8, there correspond the SDQ values for I of 1, 5, 7, 2, 4, and 8. I cannot have an SDQ of 3, 6, or 9. We also note that $SDQ[SDQ(p) + SDQ(I)] = 2, 7$. To get an SDQ of 1, we use (all SDQ): 1x1, 2x5, 5x2, 4x7, 7x4, 8x8,

or: 1x1; 2x5; 4x7; 8x8. These results are seen in Table 1.

**TABLE 1**

| SDQ(p) x SDQ(I) = 1 | | | | | | |
|---|---|---|---|---|---|---|
| SDQ(p) | 1 | 2 | 4 | 5 | 7 | 8 |
| SDQ(I) | 1 | 5 | 7 | 2 | 4 | 8 |
| Sum | 2 | 7 | 2 | 7 | 2 | 7 |

To recap we have two conditions on the SDQ of the prime number and on the integer. The product of the SDQ of the prime number and the integer must equal 1, while the SDQ of the sum of the SDQ=s must equal 2, or 7.

Consider the $((n-1)! + 1)/n$, being a prime number generator, of n, if the result is an integer. This is the condition that n be a prime number, Table 2.

## TABLE 2: PRIME NUMBER CONDITION

| p | (p-1)! + 1 | ((p-1)! + 1)/p | prime? |
|---|---|---|---|
| 0 | 2 | infinity | no |
| 1 | 2 | 2 | yes |
| 2 | 2 | 1 | yes |
| 3 | 3 | 1 | yes |
| 4 | 7 | 7/4 | no |
| 5 | 25 | 5 | yes |
| 6 | 121 | 121/6 | no |
| 7 | 721 | 103 | yes |
| 8 | 5041 | 5041/8 | no |

## A LARGE PRIME NUMBER

Consider $2^{3217} - 1$, one of the largest known prime numbers. Let us test it using the SDQ, which for exponents of the base 2, yields a 6 term repeating sequence 1, 2, 4, 8, 7, 5. See following table of reduced SDQ of exponentials. A remainder of 0 produces an SDQ of 1, a remainder of 1 produces an SDQ of 2, etc. Now 3217/6 = 536, with a remainder of 1, which therefore yields an SDQ of 2. Thus, $SDQ(2^{3217}) = 2$, and $SDQ(2^{3217}) - 1 = 1$. This is an <u>incorrectness</u> test of the calculated result. Thus, if we add all of the digits of the calculated result of $2^{3217} -$ 1, the sum of the digits, when reduced to a single digit, should be a 1. If this does not produce a 1, then the calculated result is wrong. If it does produce a result of 1, then the calculated result could still be incorrect. That is why this is an incorrectness test.

## TABLE 3: REDUCED SDQ OF EXPONENTIALS

| n | 0 | 1 | 2 | 3 | 4 | 5 | 6 | 7 | 8 | 9 | 10 | 11 | 12 |
|---|---|---|---|---|---|---|---|---|---|---|----|----|----|
| x |   |   |   |   |   |   |   |   |   |   |    |    |    |
| 1 | 1 | 1 | 1 | 1 | 1 | 1 | 1 | 1 | 1 | 1 | 1 | 1 | 1 |
| 2 | 1 | 2 | 4 | 8 | 7 | 5 | 1 | 2 | 4 | 8 | 7 | 5 | 1 |
| 3 | 1 | 3 | 9 | 9 | 9 | 9 | 9 | 9 | 9 | 9 | 9 | 9 | 9 |
| 4 | 1 | 4 | 7 | 1 | 4 | 7 | 1 | 4 | 7 | 1 | 4 | 7 | 1 |
| 5 | 1 | 5 | 7 | 8 | 4 | 2 | 1 | 5 | 7 | 8 | 4 | 2 | 1 |
| 6 | 1 | 6 | 9 | 9 | 9 | 9 | 9 | 9 | 9 | 9 | 9 | 9 | 9 |
| 7 | 1 | 7 | 4 | 1 | 7 | 4 | 1 | 7 | 4 | 1 | 7 | 4 | 1 |
| 8 | 1 | 8 | 1 | 8 | 1 | 8 | 1 | 8 | 1 | 8 | 1 | 8 | 1 |
| 9 | 1 | 9 | 9 | 9 | 9 | 9 | 9 | 9 | 9 | 9 | 9 | 9 | 9 |

## WILSON'S THEOREM AND A LARGE PRIME NUMBER

Then, $SDQ(2^{3217} - 1) = SDQ(2^{3217}) - SDQ(1) = 2 - 1 = 1$. This does not violate the prime number possibility (not a 3, 6 or 9) and indicates that the SDQ of the integer is also a 1. Thus $n = 2^{3217} - 1$, satisfies the $((n-1)! + 1)/n = I$, and $SDQ(I)$ here equals 1. Again, we thus have the possible values: 1x1, 2x5, 4x7, 8x8, four possible combinations, including 5x2, and 7x4. It is clear that there cannot be an SDQ of: 3, 6 or 9.

## FERMAT'S THEOREM

If p is any prime which does not divide the integer a, then
$$a^{p-1} = 1 \pmod p. \text{ Or } a^{p-1} -1 = 0 \pmod p.$$
a) suppose $p = 3$, and $a = 5$, then $5^{3-1} = 25 = 1 \pmod 3$, indeed $25 = 3x8 + 1$.
b) suppose $p = 5$, and $a = 3$, then $3^{5-1} = 81 = 1 \pmod 5$, indeed $81 = 5x16 + 1$.
c) suppose $p = 23$ and $a = 5$, then $5^2 = 2$, $5^4 = 4$, $5^8 = 16 = -7$, $5^{16} = 49 = 3$, $5^{20} = 12$, and finally, $5^{22} = 1$, all $\pmod{23}$. Thus, it is easy to use if $p > a$, for then p cannot possibly divide a. (We are working with positive integers).

## FERMAT'S TWO SQUARE THEOREM

The primes, if we ignore the special class 2, may be be arranged in two classes of primes, 4n+1 and 4n-1. the first class leaves a remainder of 1, when divided by 4. The second class leaves a remainder of 3, when divided by 4. Again, class one: 4n+1 leaves a remainder of 1 when divided by 4. [4n+1 = 1

149

(mod 4)]. Class two: 4n-1 leaves a remainder of -1 or 3 when divided by 4. [4n-1 = -1 = 3  (mod 4)].

Class one: 5, 13, 17, 29, 37,  41, ... , 4n+1    remainder +1. Class two: 3, 7, 11, 19, 23, 31, ... ,   4n-1   remainder -1 or +3.

All the primes of the first class, and none of the second, can be expressed as the sum of two integral squares: thus, $5 = 1^2 + 2^2$, $13 = 2^2 + 3^2$, $17 = 1^2 + 4^2$,  $29 = 2^2 + 5^2$. Note that the SDQ of the numbers that are squared are only 1, 4, 7, and 9. Thus, the SDQ of the final number is the sum of the SDQ of the squares, Table 4.

**TABLE 4: SDQ OF SQUARED NUMBERS**

|   | 1 | 4 | 7 | 9 |
|---|---|---|---|---|
| 1 | 2 | 5 | 8 | 1 |
| 4 | 5 | 8 | 2 | 4 |
| 7 | 8 | 2 | 5 | 7 |
| 9 | 1 | 4 | 7 | 9 |

Class one primes, 4n+1, are: 5, 13, 17, 29, 37, 41, 53, 61, 73, 89, 97, 101, 109, 113, ... . N.B. SDQ(4n+1) = 5, 4, 8, 2, 1, 5, 8, 7, 1, 8, 7, 2, 1, 5, ... .

Prime numbers do not end in 2, 4, 5, 6, or in 8, (they must be odd other than the first number 2) and do not have an SDQ of 3, 6 or 9. Thus, they can only end in 1, 3, 7, and 9. A prime can always be written as the product of primes. Other than the first prime of 2, it can always be written as the sum of an even and an odd number, thus, odd + even = prime  (excluding 2). Since the prime cannot end in a 5, it must not be composed of two numbers that result in a sum whose final place is 5, thus the forbidden odd and even numbers are: end in 2 and end in 3; end in 4 and end in 1, Table 5.

**TABLE 5: SDQ OF ADDITION**

|   | 1 | 2 | 3 | 4 | 5 | 6 | 7 | 8 | 9 |
|---|---|---|---|---|---|---|---|---|---|
| 1 | 2 | 3 | 4 | 5 | 6 | 7 | 8 | 9 | 1 |
| 2 | 3 | 4 | 5 | 6 | 7 | 8 | 9 | 1 | 2 |
| 3 | 4 | 5 | 6 | 7 | 8 | 9 | 1 | 2 | 3 |
| 4 | 5 | 6 | 7 | 8 | 9 | 1 | 2 | 3 | 4 |
| 5 | 6 | 7 | 8 | 9 | 1 | 2 | 3 | 4 | 5 |
| 6 | 7 | 8 | 9 | 1 | 2 | 3 | 4 | 5 | 6 |
| 7 | 8 | 9 | 1 | 2 | 3 | 4 | 5 | 6 | 7 |
| 8 | 9 | 1 | 2 | 3 | 4 | 5 | 6 | 7 | 8 |
| 9 | 1 | 2 | 3 | 4 | 5 | 6 | 7 | 8 | 9 |

**TABLE 6: SDQ OF MULTIPLICATION**

|   | 1 | 2 | 3 | 4 | 5 | 6 | 7 | 8 | 9 |
|---|---|---|---|---|---|---|---|---|---|
| 1 | 1 | 2 | 3 | 4 | 5 | 6 | 7 | 8 | 9 |
| 2 | 2 | 4 | 6 | 8 | 1 | 3 | 5 | 7 | 9 |
| 3 | 3 | 6 | 9 | 3 | 6 | 9 | 3 | 6 | 9 |
| 4 | 4 | 8 | 3 | 7 | 2 | 6 | 1 | 5 | 9 |
| 5 | 5 | 1 | 6 | 2 | 7 | 3 | 8 | 4 | 9 |
| 6 | 6 | 3 | 9 | 6 | 3 | 9 | 6 | 3 | 9 |
| 7 | 7 | 5 | 3 | 1 | 8 | 6 | 4 | 2 | 9 |
| 8 | 8 | 7 | 6 | 5 | 4 | 3 | 2 | 1 | 9 |
| 9 | 9 | 9 | 9 | 9 | 9 | 9 | 9 | 9 | 9 |

Table 6 can be analyzed as follows:

1- There are 6 each of the numbers 1, 2, 4, 5, 7, and 8. There are 12 each of number 3, and number 6. There are 21 number 9. Thus, 36 + 24 + 21 = 81 numbers.

2- The SDQ values of 3, 6, and 9 have been shaded to indicate that they cannot be the result of a prime number.

3- There are 9 squares indicated by the colored outlines, and there are 4 different squares, the center square is unique, there are two each of the corner squares, and there are 4 squares forming a cross.

4- Each of the 9 squares adds to an SDQ of 9.

5- Viewing each of the 9 squares as a determinant, the value of the SDQ of each product is 4, and the difference in the products, called the determinant value, is 0.

The center square is unique in that there is no other like it, and its cell contents total: $7 + 2 + 7 + 2 == 18 ==> 9$. The 7 x 7 product is 49 ==> 4, and the 2 x 2 product is also 4, so that the difference in these two products, called the determinant value, (SDQ) is 0.

| 7 | 2 |
|---|---|
| 2 | 7 |

Consider the upper left hand square

| 1 | 2 |
|---|---|
| 2 | 4 |

We notice that 1 x 4 = 4, and 2 x 2 = 4, so that the determinant equals 0. Also, the sum of the cell contents is 9. The square is also symmetric about rotation; clockwise being 1, 2, 4, 2, and counterclockwise being 1, 2, 4, 2, the same.

The lower right hand square is

| 4 | 2 |
|---|---|
| 2 | 1 |

This is like the upper left hand square with the 1 and 4 changing places.

The upper right hand square is

| 7 | 8 |
|---|---|
| 5 | 7 |

while the lower left hand square is

| 7 | 5 |
|---|---|
| 8 | 7 |

These two squares are the same except for the 8 and 5 changing places. We notice that 7 x 7 = 49 => 4, and 8 x 5 = 40 => 4, so that the determinant equals 0. Also the sum of the cells, 7+5+7+8 = 27 => 9. This square is also symmetric about rotation, being the same clockwise as counterclockwise. 7, 5, 7, 8 and 7, 5, 7, 8 if we start now from the lower right 7.

The last set of 4 squares are every combination of 1, 4, 5, and 8.

|   |   |
|---|---|
| 4 | 5 |
| 8 | 1 |

|   |   |
|---|---|
| 4 | 8 |
| 5 | 1 |

|   |   |
|---|---|
| 1 | 5 |
| 8 | 4 |

|   |   |
|---|---|
| 1 | 8 |
| 5 | 4 |

Again, each adds to an SDQ of 9, and each product is 4 and each determinant value is 0.

2, 3, 5, 7, 11, 13, 17, 19, 23, 29, 31, 37, 41, 43, 47, 53, 59, 61, 67, 71, 73, 79, 83, 89, 97, 101, 103

2, 5, 2, 5, 7, 2, 7, 2, 2, 5, 5, 2, 7, 2, 2, 7, 2, 7, Their consecutive products are: 2, 10, 20, 100, 700, 11

11, 253, 7337, 300817, 12935131, with SDQ values of 2, 1, 2, 1, 7, 5, 8, 7, 5, 1, 5, 7, Their number qualities are:

0, 1, 2, 3, 5, 7, 2, 4, 8, 1, 5, 2, 4, 1, 5, 7, 2, 8, 5, 7, 4, 8, 1,

7, 2, 8, 7,   2, 4, 8, 1, 5, 1, 5, 2, 4, 5, 7, 4, 1,

5, 2, 8, 1, 2, 4, 8, 1,

**MERSENNE NUMBERS**

Prime numbers, other than 2, are always odd, of the form $2n + 1$, or $2n - 1$, where n takes on all integer values 0, 1, 2, 3, ... The series $2n+1$ then has the values; 2, 3, 5, 9, 17, 33, ... . The SDQ$(2n) = 1, 2, 4, 8, 7$, or 5. Thus, if we use the former, then the SDQ$(2n + 1) = 2, 3, 5, 9, 8, 6$. Thus the SDQ have six possible values. If we use the latter form, then since SDQ$(1 - 1) = 9$, we get the six possible values; 9, 1, 3, 7, 6, or 4. We note that the values 3, 6, and 9 appear in both representations.

$M_p = 2^p - 1$, where p = 2,3,5,7,11,...,257, ...,  p is a prime number.

Consider the values of p where $M_p$ is prime. and: p = 2,3,5,7,13,17,19,31,61,127

SDQ$(2^p) = 1\ 2\ 4\ 8\ 7\ 5\ 1\ 2\ 4\ 8\ \ 7\ 5 ...$

    $p = 0\ 1\ 2\ 3\ 4\ 5\ 6\ 7\ 8\ 9\ 10\ 11 ...$

Consider Table 7, where we list p, $M_p$, $2^p$, r, SDQ$(2^p)$ and SDQ$(2^p-1)$. The remainder r, is found by dividing the value of p by 6, and using the sequential scheme above. Thus r = 0 yields 1, r = 1 yields 2, r = 2 yields 4, r = 3 yields 8, r = 4 yields 7, and r = 5 yields 5.

**TABLE 7: SDQ($M_p$) MERSENNE NUMBERS ($2^p-1$)**

| p | $2^p$ | r | SDQ($2^p$) | $2^p-1$ | SDQ($2^p-1$) |
|---|---|---|---|---|---|
| 2 | 4 | | 4 | 3 | 3 |
| 3 | 8 | | 8 | 7 | 7 |
| 5 | 32 | | 5 | 31 | 4 |
| 7 | 128 | 1 | 2 | 127 | 1 |
| 11 | 2048 | 5 | 5 | 2047 | 4 |
| 13 | 8192 | 1 | 2 | 8191 | 1 |
| 17 | 131072 | 5 | 5 | 131071 | 4 |
| 19 | 524288 | 1 | 2 | 524287 | 1 |
| 23 | 8388608 | 5 | 5 | 8388607 | 4 |
| 29 | 536870912 | 5 | 5 | 536870911 | 4 |
| 31 | 2147483648 | 1 | 2 | 2147483647 | 1 |

With the exception of the first prime number 2, all of the SDQ($2^p-1$), are 1, 4, and 7.

**TABLE 8: SDQ(MP) MERSENNE NUMBERS** ($2p+1$)

| p | $2^p$ | | r | SDQ($2^p$) | $2p+1$ | SDQ($2p+1$) |
|---|---|---|---|---|---|---|
| 2 | 4 | | | 4 | 5 | 5 |
| 3 | 8 | | | 8 | 9 | 9 |
| 5 | 32 | | | 5 | 33 | 6 |
| 7 | 128 | | 1 | 2 | 129 | 3 |
| 11 | 2048 | | 5 | 5 | 2049 | 6 |
| 13 | 8192 | | 1 | 2 | 8193 | 3 |
| 17 | 131072 | | 5 | 5 | 131073 | 6 |
| 19 | 524288 | | 1 | 2 | 524289 | 3 |
| 23 | 8388608 | | 5 | 5 | 8388609 | 6 |
| 29 | 536870912 | | 5 | 5 | 536870913 | 6 |
| 31 | 2147483648 | | 1 | 2 | 2147483649 | 3 |

From Table 8 we note that with the exception of the first prime number 2, all of the SDQ($2p+1$), are 3, 6, and 9. By Rule 1 these cannot be prime numbers.

## FERMAT'S NUMBERS

Fermat's numbers are defined by $F_n = 2^{\wedge}2^n + 1$, where $2^{\wedge}2^n$ means 2 raised to the $2^n$ power, and where n = 0, 1, 2, 3, 4, ... . Evaluating Fermat's numbers we get $F_0 = 1$, $F_1 = 5$, $F_2 = 17$, $F_3 = 257$, $F_4 = 65537$, and $F_6 = 18{,}446{,}744{,}073{,}709{,}551{,}616$. Since prime numbers other than 2, are always odd, we may consider them of the form $2^n + 1$, or $2^n - 1$. The SDQ($2^n$) = 1, 2, 4, 8, 7, 5, a six value repeating sequence. Thus, if we use $2^n + 1$, then SDQ($2^n + 1$) = 2, 3, 5, 9, 8, 6. If we use $2^n - 1$, then since SDQ($1 - 1$) = 9, we get six possible values; 9, 1, 3, 7, 6, 4, ... We note that the values 3, 6, and 9 appear in both forms. Note that $2^n$ yields 1, 2, 4, 8, 16, 32, ..., while SDQ($2^n$) is a six term repeating sequence, mod 6, of the form $2^{6m}$, $2^{6m+1}$, $2^{6m+2}$, $2^{6m+3}$, $2^{6m+4}$, $2^{6m+5}$, where m = 0, 1, 2, 3, 4, ... . Thus $2^{\wedge}2^n$ produces $2^1$, $2^2$, $2^4$, $2^8$, $2^{16}$, $2^{32}$, $2^{64}$, ... . Finally, SDQ($F_n$) = 1, 5, 8, 5, 8, 5, 8, ... , see Table 9 for the SDQ of Fermat=s numbers.

**TABLE 9: SDQ OF FERMAT'S NUMBERS**

| n | $2^n$ | $2^{2^n}$ | $F_n$ | SDQ($2^{2^n}$) | SDQ($F_n$) |
|---|---|---|---|---|---|
| 0 | 1 | 2 | 1 | 2 | 1 |
| 1 | 2 | 4 | 5 | 4 | 5 |
| 2 | 4 | 16 | 17 | 7 | 8 |
| 3 | 8 | 256 | 257 | 4 | 5 |
| 4 | 16 | 65,536 | 65,537 | 7 | 8 |
| 5 | 32 | 4,294,967,296 | 4,294,967,297 | 4 | 5 |
| 6 | 64 | 18446744073709551616 | 18446744073709551617 | 7 | 8 |

Now F(5) = 4,294,967,197, and SDQ(F(5)) = 5. Thus if F(5) is not prime and is composed of the product of primes, then as has been shown: 4,294,967,197 = 641 x 6700417; where SDQ(641) = 2 and SDQ(6700417) = 7, thus 2x7= 14==5, and it checks. Since no prime has a SDQ of 3, 6, or 9; then as a prime is the product of primes, non of the factor primes can have a SDQ of 3,6 or 9. Thus the pattern of SDQ($F_n$ )where n ≥ 1 = 5, 8, 5, 8, ... . Thus, the SDQ pattern of Fermat's numbers is simple and cyclic.

**Rule 6: The SDQ of Fermat's numbers alternate between 5 and 8, for n ≥ 1. $F_{2n}$ = 8 (mod 9), and $F_{2n-1}$ = 5 (mod 9), for n ≥ 1.**

## PERFECT NUMBERS

Perfect numbers are equal to the sum of their factors, including 1 and excluding themselves. The factors including 1 and excluding the number itself is called the aliquot parts. Thus, 6 has factors of 1, 2, and 3, and the sum of the aliquot parts 1+2+3 = 6. The next higher perfect number is 28, which has factors of 1, 2, 4, 7, and 14, and the sum of the aliquot parts 1+2+4+7+14 = 28. The next perfect number is 496.

We can generate the perfect numbers in the following manner. Start with the $2^n$ series; 1, 2, 4, 8 16, 32, 64, 128, 256, ... . Add the terms until a prime number is reached. Thus, 1 + 2 = 3 a prime number. Then multiply that prime number by the last term used, in this case, 2. This yields 3 x 2 = 6 a perfect number. The next perfect number is obtained by adding 1+2+4 = 7 a prime number, and then multiplying it by the last term 4, to get 28. The next perfect number is obtained by adding 1+2+4+8+16 = 31, and then multiplying this prime number by 16, the last term and we get 31 x 16 = 496, a perfect number.

Notice that when we add the terms of the $2^n$ series, the sum is always 1 less than the next term. Thus, the next term of 16 is 32, and 1 less that 32 is 31, the sum of the preceding terms.

## SDQ OF PERFECT NUMBERS

The $2^n$ series is: 1, 2, 4, 8, 16, 32, 64, 128, 256, 512, ... . The SDQ of these terms are: 1, 2, 4, 8, 7, 5; 1, 2, 4, 6, 7, 5; ... . The sum of the terms of the $2^n$ series are: 1, 3, 7, 15, 31, 63, 127, 255, 511, ... always one less than the next term in the $2^n$ series. Now the SDQ of these sums are: 1, 3, 7, 6, 4, 9; 1, 3, 7, ... .

The perfect numbers are formed from the product of these two series, as noted before. Thus, the SDQ values (SDQ of a product is the product of the SDQ=s) are obtained from the SDQ of the products of (1, 2, 4, 8, 7, 5) x (1, 3, 7, 6, 4, 9) = 1x1, 2x3, 4x7, 8x6, 7x4, and 5x9, since we must follow term by term, the order of each series. These have SDQ values of 1, 6, 1, 3, 1, and 9. Thus, the perfect numbers can only have these SDQ values: 1, 3, 6 and 9. We can go one step farther, since no prime number can have an SDQ of 3, 6, or 9, this eliminates the second, fourth and sixth term from the sum, and the only remaining SDQ values are 1. Consequently, all perfect numbers must have an SDQ of 1. The ONLY exception is the first perfect number 6, since it is a single digit and does not have to be reduced to a SDQ. Thus, after the first 6, all of the perfect numbers have an SDQ of 1.

A more mathematical way of getting the SDQ results is by noting the $\sum(2^n) = 2^{i+1} - 1$, when the summation goes from n=0 to i≥2. Then we get the resulting perfect number by multiplying this number (if it is a prime) by the last term $2^i$. Thus, all perfect numbers are generated by $2^i (2^{i+1} - 1)$ when the value in parentheses is a prime.

## TRIPLE PRIME NUMBERS

Triple prime numbers exist and the first two are:

**599 =>23=>5 , amd 59,999 =>41.=>5.**

[FAR} **Investigate the existence of higher order multiple primes**

SUMMARY OF RULES

Primary Rule: p = ± 1 (mod 6).  Every prime number p ≥ 5, when divided by 6, has a remainder of 1 or 5.

Prime Number Generator

To generate all the primes ≥ 5: let $p_n$ be the primes ≥ 5 in increasing order, $p_1 = 5$, $p_2 = 7$, $p_3 = 11$, $p_4 = 13$, ... . Let $q_i$ be the value of the $i^{th}$ term of the 2n terms generated by (6n ± 1). Let N = max(6n+1).

> 1- write down the (6n ± 1) terms in increasing order, where n = 1, 2, 3, ...
>
> 2- cancel out all 5(5, 7, 11, 13, 17, 19, 23, 25, 29, 31, 35, ..., $q_j$) terms, where $5q_j ≤ N$
>
> 3- cancel out all 7(7, 11, 13, ..., $q_k$) terms that remain, where
>     $7q_k ≤$               N
>
> 4- continue to cancel out all $p_n(p_n, q_s, q_{s+1}, ..., q_l)$ terms, where $p_nq_l ≤ N$
>
> 5- the final cancellation will be $p_m(p_m)$, where $p_mp_m ≤ N$.
>
> 6- the remaining terms are all the primes ≥ 5.

Rule 1: For all prime numbers or incomposite numbers greater than 3, there are none with SDQ values of 3, 6, or 9.

Rule 2: The SDQ values of the prime prime numbers are 2, 5, or 7.  A prime prime number is a prime number whose SDQ are all primes.

Rule 3: The SDQ of the product of consecutive primes (excluding the first whose value is 2), are either 3 or 6.

Rule 4: The SDQ of the prime numbers formed from the product of consecutive primes plus 1, is either 4 or 7.

Rule 5: SDQ(n!) = 9, where n ≥ 6.

Rule 6: $F_{2n} = 8$ (mod 9), and $F_{2n-1} = 5$ (mod 9), for n ≥ 1. The SDQ of Fermat‚s numbers alternate between 5 and 8, for n ≥ 1, (8 for n even and 5 for n odd).

Theorem 276. If $2^{n+1} - 1$ is prime, then $2^n(2^{n+1} - 1)$ is perfect.

Theorem 277. Any even perfect number is a Euclid number, that is to say of the form $2^n(2^{n+1} - 1)$, where $2^{n+1} - 1$ is prime.

Rule 7: SDQ$[2^n(2^{n+1}-1)]$ = 6, 1, 3, 1, 9, 1; ..., for n = 1, 2, 3, 4, 5, 6, ... .

Rule 7a: It follows from rule 7, that SDQ$[2^n(2^{n+1}-1)]$ = 1 for n even.

**GENERAL FIBONACCI-LIKE SERIES**

a, b, a + b, a + 2b, 2a + 3b, 3a + 5b, 5a + 8b, 8a + 13b, 13a + 21b, 21a + 34b, 34a + 55b, 55a + 89b, 89a + 144b, 144a + 233b, 233a + 377b, 377a + 610b, 610a + 987b, 987a + 1597b, 1597a + 2584b, 2584a + 4181b, 4181a + 6765b, 6765a + 10946b, 10946a + 17711b, 17711a + 28657b; 28657a + 46368b, 46368a + 75025b, 75025a + 121393b, 121393a + 196418b, ... .

## SDQ OF GENERAL FIBONACCI-LIKE SERIES

Since the SDQ of a product equals the product of the SDQ, we take the SDQ of each of the coefficients of a and b. This equals: a, b, a + b, a + 2b, 2a + 3b, 3a + 5b, 5a + 8b, 8a + 4b, 4a + 3b, 3a + 7b, 7a + 1b, 1a + 8b, 8a + 9b, 9a + 8b, 8a + 8b, 8a + 7b, 7a + 6b, 6a + 4b, 4a + 1b, 1a + 5b, 5a + 6b, 6a + 2b, 2a + 8b, 8a + 1b; 1a + 9b, 9a + 1b, 1a + 1b, 1a + 2b, ... . Since SDQ(1a + 9b) ==> SDQ a, and SDQ(9a + 1b) ==> SDQ b, we see that every Fibonacci-like series has a period of 24 terms.

Note that SDQ 24 ==> 6.

# Chapter 10

## Zero

The number 1 is excluded from the class of prime numbers so that all numbers can be uniquely ordered as products of prime numbers. The number 1 has qualities as a monad that makes it unique and indivisible. For as many times as 1 is divided, it still produces 1. One pie when divided becomes so many of 1 piece of pie, so that the quality of 1 still exists. As a quality, 1, as the monad, is a conceptual archetype that is intangible and that remains unique unto itself.

The number zero likewise has many qualities that are unique unto itself. In the strict mathematical sense it is not considered a positive number, and like the number 1 is not considered a prime number. It represents the absence of any object that may be counted. In set theory it represents the empty set containing no elements. In shape, it has three parts, the inner finite part, the outer semi infinite part, and the circular boundary. As nothing, or rather no-thing, it can refer to that which is not-a-thing, and when a physical object is considered some thing, then zero refers to a class other than that of a physical object. It might refer to elements or members of the unseen world, or possibly to that which has no mass. In order to specify its range of applications we need to know the contexts in which it is used. This last sentence cannot be stressed too strongly.

In mathematics, the number zero is neither positive or negative, but can be considered as either, neither or both. All equations can be cast into a form where they equal zero. In that sense zero is the result of every possible equation representing every possible action.

In physics there are various types of quantities that are describable by equations. There are scalar quantities that have an amount or magnitude but no direction, such as speed and mass. There are vector quantities that have both magnitude and direction, such as velocity and weight. There are additionally, more complicated quantities such as matrices and tensors. The point is that no matter what the quantity, or equation determining its function, all can be equated to zero. In a sense, zero contains all possibilities, including no possibility.

Consider a region of space containing two waves that exhibit total destructive interference, in other words, that cancel each other. Is this space that contains two nullifying waves the same as the same space devoid of any waves? Are all zeros alike?

**ARE ALL ZEROS ALIKE?**

Zero, in a number sense, separates or divides the set of all numbers, positive and negative. Thus the number line looks as follows: ... -5, -4, -3, -2, -1, 0, +1, +2, +3, +4, +5, ... . Any number to the right of any other number is the larger of the two. If we add the set of all numbers that are the same distance from zero, we still get zero. Thus, -1 and +1 total 0, -2 and +2 total zero. An interesting question arises, does the zero resulting from 1 - 1 = 0, and the zero in 2 - 2 = 0, have the same significance? Is a business having accounts receivable (amount of money received) and accounts payable (amount it is liable for) of $2000; the same as a business with accounts receivable and payable of $10,000,000? Clearly not, the former is a smaller business value than the latter. Mathematically their balanced books are the same, but in actuality they are not. Something is missing in zero applications.

# The Qualities of Zero
[FAR]
Weighting - as per Craig

The ancients considered fractions to be related to the gods. The duality expression "As Above So Below" is perfectly suited to the numerator and denominator of fractions. All numbers are unchanged when divided by "1" and equally so for zero written as a fraction 0/1. Mathematically, 0/1 and 0/2 and 0/n are all the same but available for a new definition to expand their applications.

**We define $0/n \equiv n - n$, where n = any quantity, vector, tensor, or equation.**

$\equiv$ means definition

For the business example, n = $2,000 and n = $10,000,000 and can now be listed as 0/$2,000 and 0/$10,000,000, applying accounting practices and indicating valuations. This now shows that their books balance, and that their valuations are: $0/\$2,000 \equiv \$2.000 - \$2,000 = 0$, and $0/\$10,000,000 \equiv \$10,000,000 - \$10,000,000 = 0$. This new definition can be easily adopted in accounting procedures.

In physics $0/(\sum F - ma)$ means that $\sum F - ma = 0$ and is mathematically the same as $0/(E - mc^2)$, and writing it as $0/(F - ma) = 0/(E - mc^2)$ shows that they are not equal in significance or application.

This new definition for Zero extends its range of application.

# The Duality Inverse
[FAR]

The inverse of 0/1 which is $1/0 = \infty$ is now also available for definition associated

with dimensionality, so that $n/0 \equiv \infty^n$, where n = dimension.

Therefore, 2/0 equals 3/0 in quantity but not in quality.

**NUMBER: TRIPART NUMBER QUALITY**

Taking our cue from the shape of zero, and its three parts; inner, outer and boundary, we consider the all numbers as having a tripart quality: a middle part, that is either 0 or 1; and two equal outer parts. Then all non negative integers can be written as:

| | | | | | |
|---|---|---|---|---|---|
| 0 = 0 0 0 | 4 = 2 0 2 | 8 = 4 0 4 | 12 = 6 0 6 | 16 = 8 0 8 | 20 = 10 0 10 |
| 1 = 0 1 0 | 5 = 2 1 2 | 9 = 4 1 4 | 13 = 6 1 6 | 17 = 8 1 8 | 21 = 10 1 10 |
| 2 = 1 0 1 | 6 = 3 0 3 | 10 = 5 0 5 | 14 = 7 0 7 | 18 = 9 0 9 | 22 = 11 0 11 |
| 3 = 1 1 1 | 7 = 3 1 3 | 11 = 5 1 5 | 15 = 7 1 7 | 19 = 9 1 9 | 23 = 11 1 11 |

etc.

Thus, we notice that all numbers have a middle part--called the balance -- which is either a 0 or 1, acting as a balance or fulcrum. If the balance has the value 0, then the tripart number is considered female; if the balance is 1, then the tripart number is said to be male.

Even numbers; may have even or odd outer parts. Thus the even number 2 = 1 0 1, has odd outer parts, while the even number 4 = 2 0 2, has even outer parts. Odd numbers, too, may have either even or odd end parts. Thus the odd number 1 = 0 1 0, has even end parts, while the number 3 = 1 0 1, has odd outer parts. Thus all tripart numbers are said to be evenly even, evenly odd, oddly even, or oddly odd.

We also notice that numbers may be grouped according to their outer parts, numbers having the same outer parts, forming pairs. Thus 0 and 1 form a pair, (both have 0 as their outer parts), 2 and 3, and so forth.

## NUMBER: TRIPART NUMBER QUALITY SEX

The tripart numbers:

| | | |
|---|---|---|
| 0 = 0 0 0 | 4 = 2 0 2 | 8 = 4 0 4 |
| 1 = 0 1 0 | 5 = 2 1 2 | 9 = 4 1 4 |
| 2 = 1 0 1 | 6 = 3 0 3 | 10 = 5 0 5 |
| 3 = 1 1 1 | 7 = 3 1 3 | 11 = 5 1 5 |

| | | |
|---|---|---|
| 12 = 6 0 6 | 16 = 8 0 8 | 20 = 10 0 10 |
| 13 = 6 1 6 | 17 = 8 1 8 | 21 = 10 1 10 |
| 14 = 7 0 7 | 18 = 9 0 9 | 22 = 11 0 11 |
| 15 = 7 1 7 | 19 = 9 1 9 | 23 = 11 1 11 |

have the following sexual qualities:

| | | |
|---|---|---|
| 0 = f f f | 4 = f f f | 8 = f f f |
| 1 = f m f | 5 = f m f | 9 = f m f |
| 2 = m f m | 6 = m f m | 10 = m f m |
| 3 = m m m | 7 = m m m | 11 = m m m |

| | | |
|---|---|---|
| 12 = f f f | 16 = f f f | 20 = f f f |
| 13 = f m f | 17 = f m f | 21 = f m f |
| 14 = m f m | 18 = m f m | 22 = m f m |
| 15 = m m m | 19 = m m m | 23 = m m m |

Thus, the sex of the number, takes on the same sex as the sex of the middle number. Writing the total number's sex in caps, we get the following:

| | | |
|---|---|---|
| F = f f f | F = f f f | F = f f f |
| M = f m f | M = f m f | M = f m f |
| F = m f m | F = m f m | F = m f m |
| M = m m m | M = m m m | M = m m m |

The numbers 1, 5, 9, 13, 17, 21, and 2, 6, 10, 14, 18, 22   are of special interest, in that these numbers form a sandwich of either a male between two females, or a female between two males.

## NUMBER: TRIPART NUMBER AVERAGE QUALITY

We may consider that this representation of all numbers by a triple division produces many combinations for each number. We now add an additional property, namely, that we want the outer numbers to average the middle number. Thus consider the number 3, which may be written: 1 1 1, or 2 1 0, or 1 0 2, or 0 3 0.

Only two of these has the property that the ends average the middle.

1 1 1, here the ends add to 2 and when divided by 2 equals 1, the middle term.

2 1 0, here also the ends 2  and 0 add to 2 and when averaged by dividing by 2 equals 1, the middle term.

1 0 2, however, has the ends add to 3 which does not average to 0, the middle term.

0 3 0, also does not have the end numbers average to the middle number.

Thus 3 = 1 1 1 = 2 1 0.

Consider the number 4, it may be written as:

0 0 4, 0 4 0, 0 1 3, 3 1 0, 3 0 1, 0 3 1, 2 0 2,

0 2 2, 1 1 2, 1 2 1, and non of these satisfies the average property.

In fact the only numbers that can have this average property, are 0 and multiples of 3.

Thus  0 = 0 0 0

   3 = 1 1, 2 1 0

   6 = 2 2 2, 3 2 1, 4 2 0

   9 = 3 3 3, 4 3 2, 5 3 1, 6 3 0

   12 = 4 4 4, 5 4 3, 6 4 2, 7 4 1, 8 4 0 = 12 = 3

   15 = 5 5 5, 6 5 4, 7 5 3, 8 5 2, 9 5 1, 10 5 0 = 15 = 6

Notice the pattern that develops. These are all multiples of 3, and there are only the numbers that reduce to 0, 3, 6, 9 that exist, four numbers in all that are balanced.

The middle number is always the actual number divided by 3. The end numbers are equidistant from the middle number and average to the middle number.

These "triangular," numbers are composed of triplicates. Notice that as we go to 9, the middle terms are now female.

## SUGGESTED RESEARCH
### Random Numbers

There are questions of the validity of computer number generators. Work with the SDQ to determine if the computer program is in the range of acceptable values.

To get an SDQ of 1, we use (all SDQ): 1x1, 2x5, 5x2, 4x7, 7x4, 8x8. Or: 1x1; 2x5; 4x7; 8x8.

[See where this belongs, if it does]

TABLE 7: <u>VALUES OF FACTORIALS</u> - <u>SDQ</u>

| n | n-1 | (n-1)! | (n-1)!/n | | |
|---|---|---|---|---|---|
| 2 | 1 | 1 | | | |
| 3 | 2 | 2 | | | |
| 4 | 3 | 6 | | | |
| 5 | 4 | 24 | | | |
| 6 | 5 | 120 | 20 | 2 | |
| 7 | 6 | 720 | | | |
| 8 | 7 | 5,040 | **630** | **9** | |
| 9 | 8 | 40,320 | **4480** | **7** | |
| 10 | 9 | 362,880 | **36288** | **9** | |
| 11 | 10 | 3,628,800 | | | |
| 12 | 11 | 39,916,800 | **3326400** | **9** | |
| 13 | 12 | 479,001,600 | | | |
| 14 | 13 | 6,227,020,800 | **444787200** | **1** | |
| 15 | 14 | 87,178,291,200 | | | |
| 16 | 15 | 1,307,674,368,000 | **81729648000** | **9** | |
| 17 | 16 | 20,922,789,888,000 | | | |
| 18 | 17 | 355,687,428,096,000 | | | |
| 19 | 18 | 6,402,373,705,728,000 | | | |
| 20 | 19 | 121,645,100,408,832,000 | | | |
| 21 | 20 | 2,432,902,008,176,640,000 | | | |
| 22 | 21 | 51,090,942,171,709,440,000 | | | |
| 23 | 22 | 1,124,000,727,777,607,680,000 | | | |
| | | | | | |

**TABLE:  PRIME, SDQ, SDQ ADD,  SUM SDQ**

| PRIME SDQ | SDQ | SDQ ADD | SUM | PRIME SDQ | SDQ | SDQ ADD | SUM |
|---|---|---|---|---|---|---|---|
| 1 | 1 | 1 | 1 | 59 | 5 | 6 | 8 |
| 2 | 2 | 3 | 3 | 61 | 7 | 1 | 6 |
| 3 | 3 | 6 | 6 | 67 | 4 | 1 | 1 |
| 5 | 5 | 6 | 2 | 71 | 8 | 9 | 9 |
| 7 | 7 | 1 | 9 | 73 | 1 | 1 | 1 |
| 11 | 2 | 3 | 2 | 79 | 7 | 1 | 8 |
| 13 | 4 | 1 | 6 | 83 | 2 | 3 | 1 |
| 17 | 8 | 9 | 5 | 89 | 8 | 9 | 9 |
| 19 | 1 | 1 | 6 | 97 | 7 | 1 | 7 |
| 23 | 5 | 6 | 2 | 101 | 2 | 3 | 9 |
| 29 | 2 | 3 | 4 | 103 | 4 | 1 | 4 |
| 31 | 4 | 1 | 8 | 107 | 8 | 9 | 3 |
| 41 | 5 | 6 | 4 | 109 | 1 | 1 | 4 |
| 43 | 7 | 1 | 2 | 113 | 5 | 6 | 9 |
| 47 | 2 | 3 | 4 | 127 | 1 | 1 | 1 |
| 53 | 8 | 9 | 3 | 131 | 5 | 6 | 6 |

Other than first and only 3, there appear to be no 3, 6, or 9 SDQ values for prime numbers. Thus, we have an instant check for these three digits.

| n | $2^n$ | r(n) | S($2^n$) | $2^{(2^n)}$ | r($2^n$) | S($2^{(2^n)}$) |
|---|---|---|---|---|---|---|
| 0 | 1 | | 1 | 2 | | 2 |
| 1 | 2 | | 2 | 4 | | **4** |
| 2 | 4 | | 4 | 16 | | 7 |
| 3 | 8 | | 8 | 256 | 2 | **4** |
| 4 | 16 | | 7 | 65536 | 4 | 7 |
| 5 | 32 | | 5 | 4294967296 | 2 | **4** |
| 6 | 64 | 0 | 1 | | 4 | 7 |
| 7 | 128 | 1 | 2 | | 2 | *4* |
| 8 | 256 | 2 | 4 | | 4 | 7 |
| 9 | 512 | 3 | 8 | | 2 | *4* |
| 10 | 1024 | 4 | 7 | | 4 | 7 |
| 11 | 2048 | 5 | 5 | | 2 | *4* |
| 12 | 4096 | 0 | 1 | | 4 | 7 |
| 13 | 8192 | 1 | 2 | | 2 | *4* |
| 14 | 16384 | 2 | 4 | | 4 | 7 |
| 15 | 32768 | 3 | 8 | | 2 | *4* |
| 16 | 65536 | 4 | 7 | | 4 | 7 |
| 17 | 131072 | 5 | 5 | | 2 | *4* |
| 18 | 262144 | 0 | 1 | | 4 | 7 |
| 19 | 524288 | 1 | 2 | | 2 | *4* |
| 20 | 1048576 | 2 | 4 | | 4 | 7 |
| | | | | | | |

To get an SDQ of 1, we use (all SDQ): 1x1, 2x5, 5x2, 4x7, 7x4, 8x8.
Or: 1x1; 2x5; 4x7; 8x8.

✳✳✳✳✳✳✳✳✳✳✳✳✳✳✳✳✳✳ תם ונשלם שבח לאל בורא עולם ✳✳✳✳✳✳✳✳✳✳✳✳✳✳✳✳✳✳✳

# APPENDIX 1

**The First 1,000 Primes**

(the 1,000th is 7919)

For more information on primes see http://primes.utm.edu/

A double prime number is also a SDQ prime, shown here in Bold

The triple prime number ***599*** in this series is in bold italic

**2** 3 5 7 11 13 17 19 **23** 29

31 37 **41** **43** 47 53 **59** **61** 67 71

73 **79** 83 89 **97** **101** 103 107 109 **113**

127 **131** **137** 139 **149** **151** 157 163 **167** **173**

179 181 **191** 193 197 199 211 **223** **227** 229

233 **239** **241** 251 **257** **263** 269 271 **277** **281**

283 **293** 307 **311** **313** **317** **331** 337 **347** **349**

**353** 359 **367** 373 379 **383** **389** 397 **401** 409

**419** **421** 431 433 **439** **443** 449 **457** **461** 463

467 **479** 487 **491** 499 503 **509** 521 523 541

**547** 557 **563** **569** 571 577 **587** 593 ***599*** 601

607 613 **617** **619** 631 **641** 643 647 **653** **659**

661 **673** **677** 683 **691** 701 **709** 719 **727** 733

739 **743** 751 757 **761** 769 773 787 **797** 809

811 **821** 823 827 829 **839** **853** **857** 859 863

877 881 883 **887** **907** **911** 919 **929** 937 **941**

**947** 953 967 971 **977** **983** 991 **997** 1009 **1013**

**1019** 1021 **1031** **1033** 1039 **1049** **1051** 1061 1063 **1069**

**1087** **1091** 1093 1097 **1103** **1109** 1117 **1123** 1129 1151

1153 **1163** 1171 **1181** 1187 **1193** 1201 **1213** **1217** 1223

**1229** **1231** 1237 **1249** 1259 1277 1279 **1283** **1289** 1291

1297 **1301** **1303** **1307** **1319** 1321 1327 1361 1367 1373

1381 1399 1409 1423 1427 1429 1433 1439 1447 1451

1453 1459 1471 1481 1483 1487 1489 1493 1499 1511

1523 1531 1543 1549 1553 1559 1567 1571 1579 1583

1597 1601 1607 1609 1613 1619 1621 1627 1637 1657

1663 1667 1669 1693 1697 1699 1709 1721 1723 1733

1741 1747 1753 1759 1777 1783 1787 1789 1801 1811

1823 1831 1847 1861 1867 1871 1873 1877 1879 1889

1901 1907 1913 1931 1933 1949 1951 1973 1979 1987

1993 1997 1999 2003 2011 2017 2027 2029 2039 2053

2063 2069 2081 2083 2087 2089 2099 2111 2113 2129

2131 2137 2141 2143 2153 2161 2179 2203 2207 2213

2221 2237 2239 2243 2251 2267 2269 2273 2281 2287

| | | | | | | | | | |
|---|---|---|---|---|---|---|---|---|---|
| 2293 | 2297 | 2309 | 2311 | 2333 | 2339 | 2341 | 2347 | 2351 | 2357 |
| 2371 | 2377 | 2381 | 2383 | 2389 | 2393 | 2399 | 2411 | 2417 | 2423 |
| 2437 | 2441 | 2447 | 2459 | 2467 | 2473 | 2477 | 2503 | 2521 | 2531 |
| 2539 | 2543 | 2549 | 2551 | 2557 | 2579 | 2591 | 2593 | 2609 | 2617 |
| 2621 | 2633 | 2647 | 2657 | 2659 | 2663 | 2671 | 2677 | 2683 | 2687 |
| 2689 | 2693 | 2699 | 2707 | 2711 | 2713 | 2719 | 2729 | 2731 | 2741 |
| 2749 | 2753 | 2767 | 2777 | 2789 | 2791 | 2797 | 2801 | 2803 | 2819 |
| 2833 | 2837 | 2843 | 2851 | 2857 | 2861 | 2879 | 2887 | 2897 | 2903 |
| 2909 | 2917 | 2927 | 2939 | 2953 | 2957 | 2963 | 2969 | 2971 | 2999 |
| 3001 | 3011 | 3019 | 3023 | 3037 | 3041 | 3049 | 3061 | 3067 | 3079 |
| 3083 | 3089 | 3109 | 3119 | 3121 | 3137 | 3163 | 3167 | 3169 | 3181 |
| 3187 | 3191 | 3203 | 3209 | 3217 | 3221 | 3229 | 3251 | 3253 | 3257 |
| 3259 | 3271 | 3299 | 3301 | 3307 | 3313 | 3319 | 3323 | 3329 | 3331 |
| 3343 | 3347 | 3359 | 3361 | 3371 | 3373 | 3389 | 3391 | 3407 | 3413 |
| 3433 | 3449 | 3457 | 3461 | 3463 | 3467 | 3469 | 3491 | 3499 | 3511 |
| 3517 | 3527 | 3529 | 3533 | 3539 | 3541 | 3547 | 3557 | 3559 | 3571 |
| 3581 | 3583 | 3593 | 3607 | 3613 | 3617 | 3623 | 3631 | 3637 | 3643 |
| 3659 | 3671 | 3673 | 3677 | 3691 | 3697 | 3701 | 3709 | 3719 | 3727 |
| 3733 | 3739 | 3761 | 3767 | 3769 | 3779 | 3793 | 3797 | 3803 | 3821 |
| 3823 | 3833 | 3847 | 3851 | 3853 | 3863 | 3877 | 3881 | 3889 | 3907 |
| 3911 | 3917 | 3919 | 3923 | 3929 | 3931 | 3943 | 3947 | 3967 | 3989 |
| 4001 | 4003 | 4007 | 4013 | 4019 | 4021 | 4027 | 4049 | 4051 | 4057 |
| 4073 | 4079 | 4091 | 4093 | 4099 | 4111 | 4127 | 4129 | 4133 | 4139 |
| 4153 | 4157 | 4159 | 4177 | 4201 | 4211 | 4217 | 4219 | 4229 | 4231 |
| 4241 | 4243 | 4253 | 4259 | 4261 | 4271 | 4273 | 4283 | 4289 | 4297 |
| 4327 | 4337 | 4339 | 4349 | 4357 | 4363 | 4373 | 4391 | 4397 | 4409 |
| 4421 | 4423 | 4441 | 4447 | 4451 | 4457 | 4463 | 4481 | 4483 | 4493 |
| 4507 | 4513 | 4517 | 4519 | 4523 | 4547 | 4549 | 4561 | 4567 | 4583 |
| 4591 | 4597 | 4603 | 4621 | 4637 | 4639 | 4643 | 4649 | 4651 | 4657 |
| 4663 | 4673 | 4679 | 4691 | 4703 | 4721 | 4723 | 4729 | 4733 | 4751 |
| 4759 | 4783 | 4787 | 4789 | 4793 | 4799 | 4801 | 4813 | 4817 | 4831 |
| 4861 | 4871 | 4877 | 4889 | 4903 | 4909 | 4919 | 4931 | 4933 | 4937 |
| 4943 | 4951 | 4957 | 4967 | 4969 | 4973 | 4987 | 4993 | 4999 | 5003 |
| 5009 | 5011 | 5021 | 5023 | 5039 | 5051 | 5059 | 5077 | 5081 | 5087 |
| 5099 | 5101 | 5107 | 5113 | 5119 | 5147 | 5153 | 5167 | 5171 | 5179 |
| 5189 | 5197 | 5209 | 5227 | 5231 | 5233 | 5237 | 5261 | 5273 | 5279 |
| 5281 | 5297 | 5303 | 5309 | 5323 | 5333 | 5347 | 5351 | 5381 | 5387 |
| 5393 | 5399 | 5407 | 5413 | 5417 | 5419 | 5431 | 5437 | 5441 | 5443 |
| 5449 | 5471 | 5477 | 5479 | 5483 | 5501 | 5503 | 5507 | 5519 | 5521 |
| 5527 | 5531 | 5557 | 5563 | 5569 | 5573 | 5581 | 5591 | 5623 | 5639 |

5641  5647  5651  5653  5657  5659  5669  5683  5689  5693
5701  5711  5717  5737  5741  5743  5749  5779  5783  5791
5801  5807  5813  5821  5827  5839  5843  5849  5851  5857
5861  5867  5869  5879  5881  5897  5903  5923  5927  5939
5953  5981  5987  6007  6011  6029  6037  6043  6047  6053
6067  6073  6079  6089  6091  6101  6113  6121  6131  6133
6143  6151  6163  6173  6197  6199  6203  6211  6217  6221
6229  6247  6257  6263  6269  6271  6277  6287  6299  6301
6311  6317  6323  6329  6337  6343  6353  6359  6361  6367
6373  6379  6389  6397  6421  6427  6449  6451  6469  6473
6481  6491  6521  6529  6547  6551  6553  6563  6569  6571
6577  6581  6599  6607  6619  6637  6653  6659  6661  6673
6679  6689  6691  6701  6703  6709  6719  6733  6737  6761
6763  6779  6781  6791  6793  6803  6823  6827  6829  6833
6841  6857  6863  6869  6871  6883  6899  6907  6911  6917
6947  6949  6959  6961  6967  6971  6977  6983  6991  6997
7001  7013  7019  7027  7039  7043  7057  7069  7079  7103
7109  7121  7127  7129  7151  7159  7177  7187  7193  7207
7211  7213  7219  7229  7237  7243  7247  7253  7283  7297
7307  7309  7321  7331  7333  7349  7351  7369  7393  7411
7417  7433  7451  7457  7459  7477  7481  7487  7489  7499
7507  7517  7523  7529  7537  7541  7547  7549  7559  7561
7573  7577  7583  7589  7591  7603  7607  7621  7639  7643
7649  7669  7673  7681  7687  7691  7699  7703  7717  7723
7727  7741  7753  7757  7759  7789  7793  7817  7823  7829
**7841**  **7853**  7867  **7873**  **7877**  7879  7883  7901  **7907**  7919

Prime numbers end in 1, 3, 7 or 9 (other than the single digits 2, 5 and 7).
Double primes have an SDQ of **2**, **5** or **7**. The 3 appears once as a single digit.
There are 80 double primes in the first 1,000 primes.

1/1170 = 0.00085470085470085470085470085470085470085470854

# Appendix 2

## MAGIC SQUARES

### BACKGROUND

The history of magic squares is uncertain, but it is known that the Chinese, Egyptians, Pythagoreans, Platonists and Arabs employed them. Used for magical and astrological purposes by Solomon, it is said that the Kabbalists used them on amulets and for prophetic divination. There are hints that they relate to entrance to the Tree of Life, and that knowledge of their methods can be employed to control the mechanisms of the universe. It is known that Benjamin Franklin, Albrecht Durer, and others used them at least for recreation.

A magic square has the property that the sum of the numbers of each row, column, and long diagonal, is the same. This sum is called the magic constant, S. The number of rows or columns in the magic square is called the order of the square, n. Magic squares are usually classified as odd, evenly-even, and even. This classification depends upon the order number, n, which can be odd, even and divisible by 2, and evenly-even divisible only by 4. The smallest magic square possible is a 3rd order square, n = 3, a 3x3 square holding 9 consecutive numbers.

The current status of magic squares confines itself to the standard magic square, where the starting number is 1, and the numbers increase consecutively. We will construct and examine magic squares using a new technique, whereby each number in each cell is reduced to a single digit. We apply the SDQ (single digit quality) to three kinds of magic squares; the "standard magic square" whose starting number is 1, and two new magic squares; the "lost magic square" whose starting number is 0, and magic squares that include both positive and negative numbers. We will study the general properties of the SDQ applied to the three kinds of magic squares.

We also define a new kind of magic square, where the product of the individual rows, columns, and long diagonals is the same. This magic "product" square is constructed using the Golden Proportion numbers a Fibonacci series. This is a periodic series when expressed as Single Digit Qualities, SDQ.

N.B. Throughout this work, we use the symbol => to indicate the reduction of a number to its single digit. Thus, 328 becomes 3+2+8 = 13, and 13 becomes 1+3 = 4. Therefore, we write the Single Digit Quality of 328, or the SDQ(328) as; 328 => 13 => 4.

## MAGIC SQUARES: QUALITATIVE DEFINITIONS

A magic square is a square usually formed from consecutive natural numbers, such that the sum of numbers in every column, every row and the long diagonals, is the same, (the magic constant). According to the late Aryeh Kaplan, "These magic squares are apparently used for a very special meditation..." Each horizontal row is referred to as a "house, "while each of the divisions of the row is known as a room.


There are as many houses as the number of rooms in each house. The rooms or cells may be filled with numbers having special significance. Thus, a house is composed of rooms each room containing a number. We will call the total number of rooms of the magic square, the square number. This parallels a matrix in mathematics, where the cells are arranged into rows and columns. A particular element of the matrix is represented by its row and column. The corner cells of a magic square are known as the vertices.

A square is additionally called "regular," if the sum of the numbers in the cells equidistant from the center, is the same. The smallest magic square, called a third order square, n = 3, is a 3 x 3 square containing nine cells.

Squares are usually analyzed according to their being of odd or even order, each having special properties. A further important category is the so called evenly-even magic square, an even order square whose order number is also divisible by 4.

Square cell numbers usually begin with the number 1. In this work, we will also examine squares that start with 0, which we call the lost magic squares. This follows from the decoded gematria of the  Sefer Yetzirah, namely, that the number corresponding to aleph is zero. We will compare the lost and standard magic squares, (those starting with the number 1), both in complete number and in their SDQ reduced number squares.

The number contained in a cell can relate to a particular quality, color, sound, and sex. Thus, a magic square can incorporate a great deal of information in both its aspect of size and content.


## BIBLICAL MAGIC SQUARE ASSOCIATIONS

We might view the complete magic square as a magic "garden" or magic "Kingdom," with its magic houses and magic rooms. The occupants of the rooms, the numbers, can be either male (odd) or female (even). The rooms can have sounds, lights, colors, and other qualities. There are different paths or passageways to the various rooms, some have (k)night moves others king moves. There is an inherent order to the layout. There are four winds that can carry one througout the kingdom, and these are reflected in the vowels. The duality of this kingdom is evidenced by the days and nights. There are four rivers that water the kingdom.

The kingdom contains gold, onyx and bdelium. [The word for onyx - shoham, shin heh mem, is a permutation of the name of Deity - Hashem, heh shin mem]. Within the kingdom it is not necessary to be clothed, since the kingdom has its borders or boundaries. Outside the kingdom, however, clothing is required for protection from the unbounded elements.

There are ten kingdoms, but the first three, 0, 1, and 2, are not available. The first available kingdom is the 3 (magic square), and other kingdoms increase until the last one, the 9. The inhabitants of these kingdoms cannot look into the squares but are two dimensional and must travel from corridor to corridor, room to room.

## KABBALISTIC MAGIC SQUARES

The square cells might contain the Hebrew letters rather than their associated numbers, lending them interesting additional properties. Houses, could be defined as the sub squares in the square. We may even pursue the study of bi magic and tri magic squares as related to the gematria.

The Kabbalistic application of magic squares identifies the rows and columns with the Hebrew letters and with the Sefirot. Since there are 10 Sefirot, 22 Hebrew letters, and 32 paths of the Tree of Life, the interesting order numbers of the squares are: 10, 22, and 32. The 10 and the 22 squares are of even  order, while the 32 square, being divisible by 4, is evenly-even.

A third order square of particular interest is one where the row and column cells are associated with the number values for the biblical expression, "ehiyeh asher ehiyeh," a Name of God. A fourth order square of interest is associated with another Name of God, "yod heh vav heh."

Note Bene: We may regard the square as looking down on a coffee table and may imagine that we may fill each cell with a third dimension. In that sense, it would appear that the square is the foundation for a house of a particular height. In fact, three dimensional magic squares exist, but we will defer to those in a later work.

The size of an individual cell has traditionally been the same for all cells. However, nothing requires that this always be the case. The individual number within the cell determines quality and sex, while cell size can be associated with aspects of special relativity. The significance of cell size will be explored in a subsequent work.

## MAGIC SQUARES: MAGIC SUM CONSTANT

It is a central property of a magic square that the sum of each row, column or diagonal, be the same. The value of this sum, S, called the magic sum constant, is given by the general formula:

$S = nA + B(n/2)(n2 - 1)$,

where; A = initial number; B = increment; n = order. The lowest order is n = 3, which is for a 3x3 square containing 9 cells. It has been standard to use a initial number of A = 1, with an increment B = 1.

We will investigate both this standard magic square and two others. A "lost magic square" where the initial number A = 0, and one where we use both positive and negative numbers, so that the total is 0.

The total sum, St, is the sum of every cell value in the square. This total sum is also clearly equal to nS, the number of rows or columns times the magic constant.

## MAGIC SQUARES: STANDARD

The magic squares in this chapter are Standard Magic Squares characterized by having 1 as their initial number, as opposed to the Lost Magic Squares of the following chapter, where 0 will be the initial number. Each chapter, however, examines the results when multiple digits are reduced to a single digit quality, called SDQ.


## QUANTITATIVE DEFINITIONS OF STANDARD MAGIC SQUARES

A standard magic square has the following properties and definitions:

Order:  n is the order number, nth order means an n by n square

Normal: a square that is formed of first n2 natural numbers. The lowest possible value of n = 3

Initial number: the starting number for filling the cells of the square. For standard squares, the starting number A = 1

Magic Constant: the magic constant, S, is the sum of all of the numbers in a single row, column or diagonal. $S = n(n2+1)/2$

Complement: the number that must be added to make the two numbers add to n2+1.

Symmetrical magic squares: are magical squares, such that any two numbers symmetric about the center, are complements.

Total sum: is the total sum, St, of all the numbers in the square. $St = n2(n2+1)/2 = nS$

Diametric sum: the diametric sum, Sd, is equal to the last term number, or largest number in the square, n2.

C = the numerical value of the center cell. $C = (n2 + 1)/2$.

The transposition property of symmetrical magic squares = any two rows or columns that are equidistant from the center of the square may be transposed and still yield a symmetrical magic square.

Sex is determined as follows:

1- the dominant sex is determined by the number being either even-female, or odd-male.

2- as the number is reduced to a single digit, each reduction is either even-female or odd-male. Thus, the number 27 is odd and dominant male, M, and it reduces to a 9 (27=>9), which is even and thus male, m. This is a secondary sexual characteristic and is not capitalized.

3- with each successive reduction to a single digit, the even or odd aspect determines the sex. A number like 19=>10=>1, has sexes Male, female and male, or Mfm.

STARTING CELL

In both systems of the standard and lost magic squares, we may start placing numbers in any but the center cell. The starting cell refers to that particular cell where the first number is placed. The starting number is the lowest number used to start the consecutive numbering within the cells of the square. In these examples of odd order squares, in both the standard and lost magic square systems, we choose the first cell to be (n+1)/2 across, and (n+3)/2 down. Thus, the first cell for a 3x3 will be (3+1)/2 = 2 across, and (3+3)/2 = 3 down, or second house, third room.

It has been found that the placement of succeeding numbers in the odd order cells has an interesting relationship to the moves of the king and the knight in chess.

For the even order squares, in both systems, we choose to place the first number in the last column of the first row cell; that is, the first house and last room...

ODD ORDER: 3X3 STANDARD MAGIC SQUARE

| 4 | 9 | 2 |
|---|---|---|
| 3 | 5 | 7 |
| 8 | 1 | 6 |

This is the most simple and lowest order magic square that can be formed. There are eight different ways to arrange these numbers and still satisfy the magic square definition. However, the number 5 must always occupy the central cell, and the number 1, can never be placed in one of the vertices. Notice that moving from the starting cell holding the 1, (called here the 1 cell), to the cell holding the 2, is a knight's move (in chess). Similarly, the cell holding the 3 is a knights move from the cell holding the 2. Then there follows king moves to the cells holding the 4, then 5, then 6. Finally, there are two knight moves to the cells holding the 8 and then the 9. Thus, there are a total of 4 knight moves and 4 king moves. If we construct another square using the SDQ values for each cell we get the same result, as all of the numbers are of a single digit. To summarize: the magic constant, S = 15, since each of the columns, rows and diagonals add to 15, the SDQ of 15 is 6, since 15 ==> 6. This may be computed using $S = n(n2+1)/2$, with n=3, so that $S = 3(9+1)/2 = 30/2 = 15$.

The total sum $St = n2(n2+1)/2$, here is $9(9+1)/2 = 45 = SDQ(9)$.

The center equals 5
The total sum minus the center equals 40 = SDQ(4)
The diametric sum, or extremes off the center, is 10
The diagonals increase by 1 and 3, (4,5,6 and 2,5,8)
The vertices are all even numbers, female

| F | M | F |
|---|---|---|
| M | M | M |
| F | M | F |

Notice that the central cross or row 2 and column 2, is male, while the vertices are all female. Each of the eight possible arrangements of the third order square has a central cross that is male, and vertices that are female. The eight are as follows:

| 4 | 9 | 2 |   | 2 | 9 | 4 |   | 8 | 1 | 6 |   | 6 | 1 | 8 |
|---|---|---|---|---|---|---|---|---|---|---|---|---|---|---|

| 3 | 5 | 7 |
|---|---|---|
| 8 | 1 | 6 |
|   |   |   |
| 4 | 3 | 8 |
| 9 | 5 | 1 |
| 2 | 7 | 6 |

| 7 | 5 | 3 |
|---|---|---|
| 6 | 1 | 8 |
|   |   |   |
| 2 | 7 | 6 |
| 9 | 5 | 1 |
| 4 | 3 | 8 |

| 3 | 5 | 7 |
|---|---|---|
| 4 | 9 | 2 |
|   |   |   |
| 8 | 3 | 4 |
| 1 | 5 | 9 |
| 6 | 7 | 2 |

| 7 | 5 | 3 |
|---|---|---|
| 2 | 9 | 4 |
|   |   |   |
| 6 | 7 | 2 |
| 1 | 5 | 9 |
| 8 | 3 | 4 |

## ODD ORDER: 5X5 STANDARD MAGIC SQUARE

| 11 | 24 | 7  | 20 | 3  |
|----|----|----|----|----|
| 4  | 12 | 25 | 8  | 16 |
| 17 | 5  | 13 | 21 | 9  |
| 10 | 18 | 1  | 14 | 22 |
| 23 | 6  | 19 | 2  | 15 |

[rem: look at other crosses and their significance]

In this 5x5 odd order standard magic square, the magic constant, S, or the sum of a row, column or diagonal equals (5*26)/2 = 65. The diagonals increase by 1 (11, 12, 13, 14, 15) and by 5 (3, 8, 13, 18, 23). The total sum adds to (25*26)/2 = 325. The center equals 13. The total sum minus the center equals 312 The extremes off the center add to 26 and equal distances from the center also equals 26. The vertices, or corner cell values 11, 3, 23, and 15, are all dominant male. The center cross, center column and center row, consists of all odd or dominant male numbers.

| Mf | Ff | M   | Ff | M  |
|----|----|-----|----|----|
| F  | Fm | Mm  | F  | Fm |
| Mf | M  | Mf  | Mm | M  |
| Fm | Fm | M   | Fm | Ff |
| Mm | F  | Mfm | F  | Mf |

## ODD ORDER: 5X5 STANDARD MAGIC SQUARE; SDQ

Now consider reducing each number in each cell to a single digit, by adding the digits until a single digit remains. Across, the cells are; 11, 24, 7, 20, and 3. Now

11 has two digits that add to 2, so that the SDQ of 11 equals 2. In like manner, SDQ(24) = 6, and SDQ(20) = 2. Notice that 7 and 3 are single digits and thus cannot be further reduced. Applying the SDQ to the entire 5x5 produces the following:

| 2 | 6 | 7 | 2 | 3 |
|---|---|---|---|---|
| 4 | 3 | 7 | 8 | 7 |
| 8 | 5 | 4 | 3 | 9 |
| 1 | 9 | 1 | 5 | 4 |
| 5 | 6 | 1 | 2 | 6 |

The magic constant, S, the sum of rows, columns or diagonals are either 20 or 29, each reducing to an SDQ => 2. We could have used the former result for the non SDQ square, namely, S = 65, and then applied the SDQ to arrive at the same value of 2. The total sum adds to 118 and SDQ(118) = 1. Or equivalently, the total sum of the former is equal to (25*26)/2 = 325 => 1, the same SDQ.

The center equals 13 => 4

The total sum minus the center equals 312 => 6

The extremes off the center add to 26 => 8

 Equal distances from the center = 26 => 8

SDQ    NUMBER OF OCCURRENCES

| SDQ | NUMBER OF OCCURRENCES |
|---|---|
| 1 | 3 |
| 2 | 3 |
| 3 | 3 |
| 4 | 3 |
| 5 | 3 |
| 6 | 3 |
| 7 | 3 |
| 8 | 2 |
| 9 | 2 |

The SDQ sexes are as follows: (note that these are the last sex of the non SDQ sexes).

| f | f | M | f | M |
|---|---|---|---|---|
| F | m | m | F | m |
| f | M | f | m | M |
| m | m | M | m | f |
| m | F | m | F | f |

## ODD ORDER: 7X7 STANDARD MAGIC SQUARE

| 22 | 47 | 16 | 41 | 10 | 35 | 4 |
|----|----|----|----|----|----|----|
| 5 | 23 | 48 | 17 | 42 | 11 | 29 |
| 30 | 6 | 24 | 49 | 18 | 36 | 12 |
| 13 | 31 | 7 | 25 | 43 | 19 | 37 |
| 38 | 14 | 32 | 1 | 26 | 44 | 20 |
| 21 | 39 | 8 | 33 | 2 | 27 | 45 |
| 46 | 15 | 40 | 9 | 34 | 3 | 28 |

The magic constant, S, equals (7*50)/2 = 175

The sum adds to (49*50)/2 = 1225

The center equals 25

The sum minus the center equals 1200

The extremes off the center add to 50

Equal distances from the center equals 50

The diagonals increase by 1 and 7

The vertices are female

Their sexes are as follows:

| Ff | Mmf | Fm | Mm | Fm | Mf | F |
|----|-----|-----|-----|-----|-----|-----|
| M | Mm | Ffm | Mf | Ff | Mf | Mmf |
| Fm | F | Ff | Mmf | Fm | Fm | Fm |
| Mf | Mf | M | Mm | Mm | Mfm | Mfm |
| Fmf | Fm | Fm | M | Ff | Ff | Ff |
| Mm | Mfm | F | Mf | F | Mm | Mm |
| Ffm | Mf | Ff | M | Fm | M | Ffm |

ODD ORDER: 7X7 STANDARD MAGIC SQUARE: SDQ

| 4 | 2 | 7 | 5 | 1 | 8 | 4 |
|---|---|---|---|---|---|---|
| 5 | 5 | 3 | 8 | 6 | 2 | 2 |
| 3 | 6 | 6 | 4 | 9 | 9 | 3 |
| 4 | 4 | 7 | 7 | 7 | 1 | 1 |
| 2 | 5 | 5 | 1 | 8 | 8 | 2 |
| 3 | 3 | 8 | 6 | 2 | 9 | 9 |
| 1 | 6 | 4 | 9 | 7 | 3 | 1 |

The columns, rows and diagonals add to 175 ==> 13 ==> 4
The sum adds to (49*50)/2 = 1225 ==> 10 ==> 1
The center equals 25 ==> 7
The sum minus the center equals 1200 ==> 3
The extremes off the center add to 50 ==> 5
Equal distances from the center equals 50 ==> 5
The diagonals increase by 1 and 7

SDQ     NUMBER OF OCCURRENCES
1                    6
2                    6
3                    6
4                    6
5                    5
6                    5

| 7 | 5 |
|---|---|
| 8 | 5 |
| 9 | 5 |

## ODD ORDER: 9X9 STANDARD MAGIC SQUARE

| 37 | 78 | 29 | 70 | 21 | 62 | 13 | 54 | 5  |
|----|----|----|----|----|----|----|----|----|
| 6  | 38 | 79 | 30 | 71 | 22 | 63 | 14 | 46 |
| 47 | 7  | 39 | 80 | 31 | 72 | 23 | 55 | 15 |
| 16 | 48 | 8  | 40 | 81 | 32 | 64 | 24 | 56 |
| 57 | 17 | 49 | 9  | 41 | 73 | 33 | 65 | 25 |
| 26 | 58 | 18 | 50 | 1  | 42 | 74 | 34 | 66 |
| 67 | 27 | 59 | 10 | 51 | 2  | 43 | 75 | 35 |
| 36 | 68 | 19 | 60 | 11 | 52 | 3  | 44 | 76 |
| 77 | 28 | 69 | 20 | 61 | 12 | 53 | 4  | 45 |

The magic constant equals 369

The total sum adds to (80*81)/2 = 3240

The center is 41

 The total sum minus the center equals 3200

The extremes off the center equals 82

Equal distances from the center equals 82

The diagonals increase by 1 and 9

The vertices are male

| 1 | 6 | 2 | 7 | 3 | 8 | 4 | 9 | 5 |
|---|---|---|---|---|---|---|---|---|
| 6 | 2 | 7 | 3 | 8 | 4 | 9 | 5 | 1 |
| 2 | 7 | 3 | 8 | 4 | 9 | 5 | 1 | 6 |
| 7 | 3 | 8 | 4 | 9 | 5 | 1 | 6 | 2 |
| 3 | 8 | 4 | 9 | 5 | 1 | 6 | 2 | 7 |
| 8 | 4 | 9 | 5 | 1 | 6 | 2 | 7 | 3 |
| 4 | 9 | 5 | 1 | 6 | 2 | 7 | 3 | 8 |
| 9 | 5 | 1 | 6 | 2 | 7 | 3 | 8 | 4 |
| 5 | 1 | 6 | 2 | 7 | 3 | 8 | 4 | 9 |

The columns, rows and diagonals add to 369 ==> 18 ==> 9
The sum adds to (80*81)/2 = 3240 ==> 9
The center is 41 ==> 5
The sum minus the center equals 3200 ==> 5
The extremes off the center equals 80 ==> 8
Equal distances from the center equals 82 ==> 10 ==> 1
The diagonals increase by 1 and 9
The vertices are male

SDQ     NUMBER OF OCCURRENCES
1                   9
2                   9
3                   9
4                   9
5                   9
6                   9
7                   9
8                   9
9                   9

If we rotate this clockwise by 45 degrees, we get an unusually simple pattern,

```
                         1
                       6   6
                     2   2   2
                   7   7   7   7
                 3   3   3   3   3
               8   8   8   8   8   8
             4   4   4   4   4   4   4
           9   9   9   9   9   9   9   9
         5   5   5   5   5   5   5   5   5
           1   1   1   1   1   1   1   1
             6   6   6   6   6   6   6
               2   2   2   2   2   2
                 7   7   7   7   7
                   3   3   3   3
                     8   8   8
                       4   4
                         9
```

The sum of the numbers equidistant from the horizontal base, 5--add to 10 => 1. Adding the numbers in vertical columns starting from either end and proceeding inward, we get; 5, 10, 15, 20, 25, 30, 35, 40, 45, 40, 35, 30, 25, 20, 15, 10, and 5. The SDQ of these values produce the side diagonals sequence, 5, 1, 6, 2, 7, 3, 8, 4, 9, 4, 8, 3, 7, 2, 6, 1, 5.

STANDARD MAGIC SQUARES: EVEN ORDER

EVEN ORDER: 4x4 (EVENLY-EVEN) STANDARD MAGIC SQUARE

| 4  | 14 | 15 | 1  |
|----|----|----|----|
| 9  | 7  | 6  | 12 |
| 5  | 11 | 10 | 8  |
| 16 | 2  | 3  | 13 |

The magic constant equals (4*17)/2 = 34
The total sum equals (16*17)/2 = 136
The center square equals 34
The sum minus the center square = 102
The vertices add to 17
The diagonals increase by 3 and 5
The vertices are male and female

A property of the even order magic squares, is that they can be divided into four sub-squares. In the 4x4 we get the following:

| 34 | 34 |
|----|----|
| 34 | 34 |

Each of the four blocks = 34.

## EVEN ORDER: 4X4 STANDARD MAGIC SQUARE: SDQ

| 4 | 5 | 6 | 1 |
|---|---|---|---|
| 9 | 7 | 6 | 3 |
| 5 | 2 | 1 | 8 |
| 7 | 2 | 3 | 4 |

The columns, rows and diagonals add to 34 => 7
The sum equals (16*17)/2 = 136 => 1
The center square equals 34 => 7
The sum minus the center square = 102 => 3
The vertices add to 17 => 8
Each of the four blocks = 34 => 7
The four blocks SDQ are 7.

| 7 | 7 |
|---|---|
| 7 | 7 |

| SDQ | NUMBER OF OCCURRENCES |
|-----|-----------------------|
| 1 | 2 |
| 2 | 2 |
| 3 | 2 |
| 4 | 2 |
| 5 | 2 |
| 6 | 2 |
| 7 | 2 |
| 8 | 1 |
| 9 | 1 |

## EVEN ORDER: 6x6 STANDARD MAGIC SQUARE

| 6 | 32 | 3 | 34 | 35 | 1 |
|---|----|---|----|----|---|
| 7 | 11 | 27 | 28 | 8 | 30 |
| 19 | 14 | 16 | 15 | 23 | 24 |
| 18 | 20 | 22 | 21 | 17 | 13 |
| 25 | 29 | 10 | 9 | 26 | 12 |
| 36 | 5 | 33 | 4 | 2 | 31 |

The columns, rows and diagonals add to 111=> 3
The sum equals (36*37)/2 = 666
The center square equals 74
The next square vertices equals 74
The outer square vertices equals 74
The sum minus the center square = 592
The vertices add to 74
The diagonals increase by 5 and 7
The vertices are male and female
Each row has 3 males and 3 females
Each column has 3 males and 3 females
Each diagonal has 3 males and 3 females
The four blocks are, reading across, 135, 198, down 198, 135.

| 135 | 198 |
|-----|-----|
| 198 | 135 |

## EVEN ORDER: 6X6 STANDARD MAGIC SQUARE: SDQ

| 6 | 5 | 3 | 7 | 8 | 1 |
|---|---|---|---|---|---|
| 7 | 2 | 9 | 1 | 8 | 3 |
| 1 | 5 | 7 | 6 | 5 | 6 |
| 9 | 2 | 4 | 3 | 8 | 4 |
| 7 | 2 | 1 | 9 | 8 | 3 |
| 9 | 5 | 6 | 4 | 2 | 4 |

The columns, rows and diagonals add to 111 => 3

The sum equals $(36*37)/2 = 666$ => 18 => 9

The center square equals 74 => 11 => 2

The next square vertices equals 74 => 11 => 2

The outer square vertices equals 74 => 11 => 2

The sum minus the center square = 592 => 16 => 7

The vertices add to 74 => 11 =>2

The four blocks are, reading across, 135, 198, down 198, 135 all => 9

The diagonals increase by 5 and 7

The vertices are male and female

Each row has 3 males and 3 females

Each column has 3 males and 3 females

Each diagonal has 3 males and 3 females

The four blocks SDQ are 9.

| 9 | 9 |
|---|---|
| 9 | 9 |

| SDQ | NUMBER OF OCCURRENCES |
|---|---|
| 1 | 4 |
| 2 | 4 |
| 3 | 4 |
| 4 | 4 |
| 5 | 4 |
| 6 | 4 |
| 7 | 4 |
| 8 | 4 |
| 9 | 4 |

EVEN ORDER: 8x8 (EVENLY-EVEN) STANDARD MAGIC SQUARE

| 8  | 58 | 62 | 5  | 4  | 59 | 63 | 1  |
|----|----|----|----|----|----|----|----|
| 49 | 15 | 51 | 12 | 13 | 54 | 10 | 56 |
| 17 | 42 | 22 | 44 | 45 | 19 | 47 | 24 |
| 32 | 39 | 35 | 29 | 28 | 38 | 34 | 25 |
| 40 | 31 | 27 | 37 | 36 | 30 | 26 | 33 |
| 41 | 18 | 46 | 20 | 21 | 43 | 23 | 48 |
| 9  | 55 | 11 | 52 | 53 | 14 | 50 | 16 |
| 64 | 2  | 6  | 61 | 60 | 3  | 7  | 57 |

The columns, rows and diagonals add to 260
The sum equals (64*65)/2 = 2080
The center square equals 130
The sum minus the center square = 1950
The vertices add to 130
Each of the four blocks = 520, 520, 520, 520
The diagonals increase by 7 and 9
The vertices are male and female

EVEN ORDER: 8X8 EVENLY-EVEN STANDARD MAGIC SQUARE: SDQ

| 8 | 4 | 8 | 5 | 4 | 5 | 9 | 1 |
|---|---|---|---|---|---|---|---|
| 4 | 6 | 6 | 3 | 4 | 9 | 1 | 2 |
| 8 | 6 | 4 | 8 | 9 | 1 | 2 | 6 |
| 5 | 3 | 8 | 2 | 1 | 2 | 7 | 7 |
| 4 | 4 | 9 | 1 | 9 | 3 | 8 | 6 |
| 5 | 9 | 1 | 2 | 3 | 7 | 5 | 3 |
| 9 | 1 | 2 | 7 | 8 | 5 | 5 | 7 |
| 1 | 2 | 6 | 7 | 6 | 3 | 7 | 3 |

The columns, rows and diagonals add to 260 => 8
The sum equals (64*65)/2 = 2080 => 10 => 1
The center square equals 130 => 4
The sum minus the center square = 1950 => 15 => 6

The vertices add to 130 => 4

Each of the four blocks = 520, 520, 520, 520 => 7

The diagonals increase by 7 and 9

The vertices are male and female

The four blocks SDQ are each 7.

| 7 | 7 |
|---|---|
| 7 | 7 |

SDQ    NUMBER OF OCCURRENCES

| 1 | 8 |
|---|---|
| 2 | 7 |
| 3 | 7 |
| 4 | 7 |
| 5 | 7 |
| 6 | 7 |
| 7 | 7 |
| 8 | 7 |
| 9 | 7 |

EVEN ORDER: 10x10 STANDARD MAGIC SQUARE

| 10 | 92 | 8 | 94 | 5 | 96 | 97 | 3 | 99 | 1 |
|---|---|---|---|---|---|---|---|---|---|
| 11 | 19 | 83 | 17 | 85 | 86 | 14 | 88 | 12 | 90 |
| 71 | 22 | 28 | 74 | 26 | 25 | 77 | 23 | 79 | 80 |
| 40 | 62 | 33 | 37 | 65 | 66 | 34 | 68 | 69 | 31 |
| 51 | 49 | 53 | 44 | 46 | 45 | 57 | 58 | 42 | 60 |
| 41 | 59 | 48 | 54 | 56 | 55 | 47 | 43 | 52 | 50 |
| 70 | 39 | 63 | 67 | 35 | 36 | 64 | 38 | 32 | 61 |
| 30 | 72 | 78 | 24 | 76 | 75 | 27 | 73 | 29 | 21 |
| 81 | 89 | 13 | 87 | 16 | 15 | 84 | 18 | 82 | 20 |
| 100 | 2 | 98 | 7 | 95 | 6 | 4 | 93 | 9 | 91 |

The columns, rows and diagonals add to 505 = 10 = 1

The sum equals (100*101)/2 = 5050 = 10 = 1

The vertices add to 198 = 18 = 9

The center square equals 202 = 4
The sum minus the center square = 4848 = 24 = 6
The diagonals increase by 9 and 11
The vertices are male and female


EVEN ORDER: 10X10 STANDARD MAGIC SQUARE: SDQ

| 1 | 2 | 8 | 4 | 5 | 6 | 7 | 3 | 9 | 1 |
|---|---|---|---|---|---|---|---|---|---|
| 2 | 1 | 2 | 8 | 4 | 5 | 5 | 7 | 3 | 9 |
| 8 | 4 | 1 | 2 | 8 | 7 | 5 | 5 | 7 | 8 |
| 4 | 8 | 6 | 1 | 2 | 3 | 7 | 5 | 6 | 4 |
| 6 | 4 | 8 | 8 | 1 | 9 | 3 | 4 | 6 | 6 |
| 5 | 5 | 3 | 9 | 2 | 1 | 2 | 7 | 7 | 5 |
| 7 | 3 | 9 | 4 | 8 | 9 | 1 | 2 | 5 | 7 |
| 3 | 9 | 6 | 6 | 4 | 3 | 9 | 1 | 2 | 3 |
| 9 | 8 | 4 | 6 | 7 | 6 | 3 | 9 | 1 | 2 |
| 1 | 2 | 8 | 7 | 5 | 6 | 4 | 3 | 9 | 1 |

The columns, rows and diagonals add to 505 => 10 => 1
The sum equals (100*101)/2 = 5050 => 10 => 1
The vertices add to 198 => 18 => 9
The center square equals 202 => 4
The sum minus the center square = 4848 => 24 => 6
The diagonals increase by 9 and 11
The vertices are male and female

| SDQ | NUMBER OF OCCURRENCES |
|-----|----------------------|
| 1 | 12 |
| 2 | 11 |
| 3 | 11 |
| 4 | 11 |
| 5 | 11 |
| 6 | 11 |
| 7 | 11 |
| 8 | 11 |
| 9 | 11 |

# Appendix 3

## MAGIC SQUARES: LOST

The new magic squares that we will construct will use the Lost Gematria letter=number system. Thus, all sequential numbering will start with an initial value of zero. Since the sum, S, of any magic square is given by: $S = nA + B(n/2)$ $(n2 -1)$, where n = order, A = initial number, and B = the increment, we see that all lost magic squares have A = 0, and thus the sum becomes, for these cases; $S = B(n/2)(n2-1)$. Also, since we will usually use an increment of B = 1, we get finally for the sum; $S = (n/2)(n2-1)$.

We will now go through a systematic study of odd and even order Lost Magic Squares.

ODD ORDER: 3x3 LOST MAGIC SQUARES

| 3 | 8 | 1 |
|---|---|---|
| 2 | 4 | 6 |
| 7 | 0 | 5 |

This is the most simple and lowest order magic square that may be formed. There are many ways of arranging these numbers and still satisfy the magic square definition. Notice that for this magic square, the SDQ square is exactly the same since all the numbers are of a single digit. A summary of this square's properties are as follows:

The magic constant, S = 12, since the columns, rows and diagonals add to 12, which has a SDQ of 3. This may be computed using
$S = n(n2-1)/2$, here n=3, so that $S = 3(9-1)/2 = 24/2 = 12$.
The total sum $St = nS = n2(n2-1)/2$, here is $9(9-1)/2 = 36 = SDQ(9)$.
The center = 4.
The sum minus the center equals 32 = SDQ(5).
The diametric sum, or extremes off the center, is 8.
The diagonals increase by 1 and 3, (3,4,5 and 1,4,7).
The vertices are all odd numbers, male.
The center cross is an even number, 4, female.

ODD ORDER: 5X5 LOST MAGIC SQUARE
First cell, (5+1)/2 = 3 across; (5+3)/2 = 4 down.

| 10 | 23 | 6 | 19 | 2 |
|----|----|----|----|----|
| 3 | 11 | 24 | 7 | 15 |
| 16 | 4 | 12 | 20 | 8 |
| 9 | 17 | 0 | 13 | 21 |
| 22 | 5 | 18 | 1 | 14 |

The columns, rows and diagonals add to 60
The sum adds to (24*25)/2 = 300
The center = 12
The sum minus the center equals 288
The extremes off the center add to 24
Equal distances from the center = 24
The diagonals increase by 1 and 5
The vertices are all female
The center cross is female

## ODD ORDER: 5X5 LOST MAGIC SQUARE: SDQ

| 1 | 5 | 6 | 1 | 2 |
|----|----|----|----|----|
| 3 | 2 | 6 | 7 | 6 |
| 7 | 4 | 3 | 2 | 8 |
| 9 | 8 | 0 | 4 | 3 |
| 4 | 5 | 9 | 1 | 5 |

The columns, rows and diagonals add to 60 ==> 6
The sum adds to (24*25)/2 = 300 ==> 3
The center = 12 ==> 3
The sum minus the center equals 288 ==> 18 ==> 9
The extremes off the center add to 24 ==> 6
Equal distances from the center = 24 ==> 6
The diagonals increase by 1 and 5

The vertices are all female
The center cross is female

| SDQ | NUMBER OF OCCURRENCES |
|-----|----------------------|
| 0   | 1                    |
| 1   | 3                    |
| 2   | 3                    |
| 3   | 3                    |
| 4   | 3                    |
| 5   | 3                    |
| 6   | 3                    |
| 7   | 2                    |
| 8   | 2                    |
| 9   | 2                    |

ODD ORDER: 7X7 LOST MAGIC SQUARE
First cell, $(7+1)/2 = 4$ across, $(7+3)/2 = 5$ down.

| 21 | 46 | 15 | 40 | 9  | 34 | 3  |
|----|----|----|----|----|----|----|
| 4  | 22 | 47 | 16 | 41 | 10 | 28 |
| 29 | 5  | 23 | 48 | 17 | 35 | 11 |
| 12 | 30 | 6  | 24 | 42 | 18 | 36 |
| 37 | 13 | 31 | 0  | 25 | 43 | 19 |
| 20 | 38 | 7  | 32 | 1  | 26 | 44 |
| 45 | 14 | 39 | 8  | 33 | 2  | 27 |

The columns, rows and diagonals add to $168 = 15 = 6$
The sum adds to $(48*49)/2 = 1176 = 15 = 6$
The center $= 24 = 6$
The sum minus the center equals $1152 = 9$
The extremes off the center add to $48 = 12 = 3$
Equal distances from the center $= 48 = 12 = 3$
The diagonals increase by 1 and 7
The vertices are all male
The center cross is female

ODD ORDER: 7X7 LOST MAGIC SQUARE: SDQ

| 3 | 1 | 6 | 4 | 9 | 7 | 3 |
|---|---|---|---|---|---|---|
| 4 | 4 | 2 | 7 | 5 | 1 | 1 |
| 2 | 5 | 5 | 3 | 8 | 8 | 2 |
| 3 | 3 | 6 | 6 | 6 | 9 | 9 |
| 1 | 4 | 4 | 0 | 7 | 7 | 1 |
| 2 | 2 | 7 | 5 | 1 | 8 | 8 |
| 9 | 5 | 3 | 8 | 6 | 2 | 9 |

The columns, rows and diagonals add to 168 ==> 15 ==> 6
The sum adds to (48*49)/2 = 1176 ==> 15 ==> 6
The center = 24 ==> 6
The sum minus the center equals 1152 ==> 9

The extremes off the center add to 48 ==> 12 ==> 3
Equal distances from the center = 48 ==> 12 ==> 3
The diagonals increase by 1 and 7
The vertices are all male
The center cross is female

| SDQ | NUMBER OF OCCURRENCES |
|---|---|
| 0 | 1 |
| 1 | 6 |
| 2 | 6 |
| 3 | 6 |
| 4 | 5 |
| 5 | 5 |
| 6 | 5 |
| 7 | 5 |
| 8 | 5 |
| 9 | 5 |

ODD ORDER: 9X9 LOST MAGIC SQUARE
First cell, (9+1)/2 = 5 across, (9+3)/2 = 6 down.

| 36 | 77 | 28 | 69 | 20 | 61 | 12 | 53 | 4  |
|----|----|----|----|----|----|----|----|----|
| 5  | 37 | 78 | 29 | 70 | 21 | 62 | 13 | 45 |
| 46 | 6  | 38 | 79 | 30 | 71 | 22 | 54 | 14 |
| 15 | 47 | 7  | 39 | 80 | 31 | 63 | 23 | 55 |
| 56 | 16 | 48 | 8  | 40 | 72 | 32 | 64 | 24 |
| 25 | 57 | 17 | 49 | 0  | 41 | 73 | 33 | 65 |
| 66 | 26 | 58 | 9  | 50 | 1  | 42 | 74 | 34 |
| 35 | 67 | 18 | 59 | 10 | 51 | 2  | 43 | 75 |
| 76 | 27 | 68 | 19 | 60 | 11 | 52 | 3  | 44 |

The columns, rows and diagonals add to 360
The sum adds to (80*81)/2 = 3240
The center is 40
The sum minus the center equals 3200
The extremes off the = 80
Equal distances from the center = 80
The diagonals increase by 1 and 9
The vertices are female
The center cross is female

ODD ORDER: 9X9 LOST MAGIC SQUARE; SDQ

| 9 | 5 | 1 | 6 | 2 | 7 | 3 | 8 | 4 |
|---|---|---|---|---|---|---|---|---|
| 5 | 1 | 6 | 2 | 7 | 3 | 8 | 4 | 9 |
| 1 | 6 | 2 | 7 | 3 | 8 | 4 | 9 | 5 |
| 6 | 2 | 7 | 3 | 8 | 4 | 9 | 5 | 1 |
| 2 | 7 | 3 | 8 | 4 | 9 | 5 | 1 | 6 |
| 7 | 3 | 8 | 4 | 0 | 5 | 1 | 6 | 2 |
| 3 | 8 | 4 | 9 | 5 | 1 | 6 | 2 | 7 |
| 8 | 4 | 9 | 5 | 1 | 6 | 2 | 7 | 3 |
| 4 | 9 | 5 | 1 | 6 | 2 | 7 | 3 | 8 |

The columns, rows and diagonals add to 360 ==> 9
The sum adds to (80*81)/2 = 3240 ==> 9
The center is 40 ==> 4
The sum minus the center equals 3200 ==> 5
The extremes off the = 80 ==> 8
Equal distances from the center = 80 ==> 8
The diagonals increase by 1 and 9
The vertices are female
The center cross is female

| SDQ | NUMBER OF OCCURRENCES |
|---|---|
| 0 | 1 |
| 1 | 9 |
| 2 | 9 |
| 3 | 9 |
| 4 | 9 |
| 5 | 9 |
| 6 | 9 |
| 7 | 9 |
| 8 | 9 |
| 9 | 8 |

If we rotate this clockwise by 45 degrees, we get an unusually simple pattern,

```
                        9
                     5     5
                  1     1     1
               6     6     6     6
            2     2     2     2     2
         7     7     7     7     7     7
      3     3     3     3     3     3     3
   8     8     8     8     8     8     8     8
4     4     4     4     4     4     4     4     4
   9     9     9     0     9     9     9     9
      5     5     5     5     5     5     5
         1     1     1     1     1     1
            6     6     6     6     6
               2     2     2     2
                  7     7     7
                     3     3
                        8
```

The numbers equidistant from the base 4, add to SDQ of 8.

## EVEN ORDER: 4X4 LOST MAGIC SQUARE

| 3  | 13 | 14 | 0  |
|----|----|----|----|
| 8  | 6  | 5  | 11 |
| 4  | 10 | 9  | 7  |
| 15 | 1  | 2  | 12 |

The columns, rows and diagonals add to 30
The sum equals (15*16)/2 = 120
The center square equals 30
The sum minus the center square = 90

The vertices add to 15
Each of the four blocks = 30
The diagonals increase by 3 and 5
The vertices are male and female

EVEN ORDER: 4X4 LOST MAGIC SQUARE: SDQ

| 3 | 4 | 5 | 0 |
|---|---|---|---|
| 8 | 6 | 5 | 2 |
| 4 | 1 | 9 | 7 |
| 6 | 1 | 2 | 3 |

The columns, rows and diagonals add to 30 ==> 3
The sum equals (15*16)/2 = 120 ==> 3
The center square equals 30 ==> 3
The sum minus the center square = 90 ==> 9
The vertices add to 15 ==> 6
Each of the four blocks = 30 ==> 3
The diagonals increase by 3 and 5
The vertices are male and female

| SDQ | NUMBER OF OCCURRENCES |
|---|---|
| 0 | 1 |
| 1 | 2 |
| 2 | 2 |
| 3 | 2 |
| 4 | 2 |
| 5 | 2 |
| 6 | 2 |
| 7 | 1 |
| 8 | 1 |
| 9 | 1 |

# EVEN ORDER: 6x6 LOST MAGIC SQUARE

| 5 | 31 | 2 | 33 | 34 | 0 |
|---|----|---|----|----|---|
| 6 | 10 | 26 | 27 | 7 | 29 |
| 18 | 13 | 15 | 14 | 22 | 23 |
| 17 | 19 | 21 | 20 | 16 | 12 |
| 24 | 28 | 9 | 8 | 25 | 11 |
| 35 | 4 | 32 | 3 | 1 | 30 |

The columns, rows and diagonals add to 105

Also 105 = 5x21 = 3x5x7 MUSIC

5+31+2+33+34+0 = 105

The sum equals (35*36)/2 = 630

The center square 15+21+14+20 = 70

The next square, 10+7+28+25 = 70

The outer square, 5+0+35+30 = 70

The sum minus the center square = 560

The vertices add to 70

The four blocks are, reading across, 126, 189, down 189, 126 all 9

The diagonals increase by 5 and 7

The vertices are male and female

Each row has 3 males and 3 females

Each column has 3 males and 3 females

Each diagonal has 3 males and 3 females

# EVEN ORDER: 6X6 LOST MAGIC SQUARE: SDQ

| 5 | 4 | 2 | 6 | 7 | 0 |
|---|---|---|---|---|---|
| 6 | 1 | 8 | 9 | 7 | 2 |
| 9 | 4 | 6 | 5 | 4 | 5 |
| 8 | 1 | 3 | 2 | 7 | 3 |
| 6 | 1 | 9 | 8 | 7 | 2 |
| 8 | 4 | 5 | 3 | 1 | 3 |

The columns, rows and diagonals add to 105 => 6

Also 105 = 5x21 = 3x5x7 MUSIC

5+31+2+33+34+0 = 105 => 6

The sum equals (35*36)/2 = 630 => 9

The center square 15+21+14+20 = 70 => 7

The next square, 10+7+28+25 = 70

The outer square, 5+0+35+30 = 70

The sum minus the center square = 560 => 11 => 2

The vertices add to 70 => 7

The four blocks are, reading across, 126, 189, down 189, 126 all 9

The diagonals increase by 5 and 7

The vertices are male and female

Each row has 3 males and 3 females

Each column has 3 males and 3 females

Each diagonal has 3 males and 3 females

| SDQ | NUMBER OF OCCURRENCES |
|-----|------------------------|
| 0   | 1                      |
| 1   | 4                      |
| 2   | 4                      |
| 3   | 4                      |
| 4   | 4                      |
| 5   | 4                      |
| 6   | 4                      |
| 7   | 4                      |
| 8   | 4                      |
| 9   | 3                      |

| 7 | 57 | 61 | 4 | 3 | 58 | 62 | 0 |
|---|----|----|---|---|----|----|---|
| 48 | 14 | 50 | 11 | 12 | 53 | 9 | 55 |
| 16 | 41 | 21 | 43 | 44 | 18 | 46 | 23 |
| 31 | 38 | 34 | 28 | 27 | 37 | 33 | 24 |
| 39 | 30 | 26 | 36 | 35 | 29 | 25 | 32 |
| 40 | 17 | 45 | 19 | 20 | 42 | 22 | 47 |
| 8 | 54 | 10 | 51 | 52 | 13 | 49 | 15 |
| 63 | 1 | 5 | 60 | 59 | 2 | 6 | 56 |

The columns, rows and diagonals add to 252
The sum equals (63*64)/2 = 2016
The center square equals 126
The sum minus the center square = 1890
The vertices add to 126
? Each of the four blocks = 504
The diagonals increase by 7 and 9
The vertices are male and female

EVEN ORDER: 8X8 LOST MAGIC SQUARE: SDQ

| 7 | 3 | 7 | 4 | 3 | 4 | 8 | 0 |
|---|---|---|---|---|---|---|---|
| 3 | 5 | 5 | 2 | 3 | 8 | 9 | 1 |
| 7 | 5 | 3 | 7 | 8 | 9 | 1 | 5 |
| 4 | 2 | 7 | 1 | 9 | 1 | 6 | 6 |
| 3 | 3 | 8 | 9 | 8 | 2 | 7 | 5 |
| 4 | 8 | 9 | 1 | 2 | 6 | 4 | 2 |
| 8 | 9 | 1 | 6 | 7 | 4 | 4 | 6 |
| 9 | 1 | 5 | 6 | 5 | 2 | 6 | 2 |

The columns, rows and diagonals add to 252 ==> 9
The sum equals (63*64)/2 = 2016 ==> 9

The center square equals 126 ==> 9

The sum minus the center square = 1890 ==> 18 ==> 9

The vertices add to 126 ==> 9

? Each of the four blocks = 504 ==> 9

The diagonals increase by 7 and 9

The vertices are male and female

| SDQ | NUMBER OF OCCURRENCES |
|-----|----------------------|
| 0   | 1                    |
| 1   | 7                    |
| 2   | 7                    |
| 3   | 7                    |
| 4   | 7                    |
| 5   | 7                    |
|     |                      |
| 6   | 7                    |
| 7   | 7                    |
| 8   | 7                    |
| 9   | 7                    |

EVEN ORDER: 10x10 LOST MAGIC SQUARE

| 9  | 91 | 7  | 93 | 4  | 95 | 96 | 2  | 98 | 0  |
|----|----|----|----|----|----|----|----|----|----|
| 10 | 18 | 82 | 16 | 84 | 85 | 13 | 87 | 11 | 89 |
| 70 | 21 | 27 | 73 | 25 | 24 | 76 | 22 | 78 | 79 |
| 39 | 61 | 32 | 36 | 64 | 65 | 33 | 67 | 68 | 30 |
| 50 | 48 | 52 | 43 | 45 | 44 | 56 | 57 | 41 | 59 |
| 40 | 58 | 47 | 53 | 55 | 54 | 46 | 42 | 51 | 49 |
| 69 | 38 | 62 | 66 | 34 | 35 | 63 | 37 | 31 | 60 |
| 29 | 71 | 77 | 23 | 75 | 74 | 26 | 72 | 28 | 20 |
| 80 | 88 | 12 | 86 | 15 | 14 | 83 | 17 | 81 | 19 |
| 99 | 1  | 97 | 6  | 94 | 5  | 3  | 92 | 8  | 90 |

The columns, rows and diagonals add to 495

The sum equals (99*100)/2 = 4950

The vertices add to 198

The center square equals 198

The sum minus the center square = 4752
The diagonals increase by 9 and 11
The vertices are male and female

## EVEN ORDER: 10X10 LOST MAGIC SQUARE; SDQ

| 9 | 1 | 7 | 3 | 4 | 5 | 6 | 2 | 8 | 0 |
|---|---|---|---|---|---|---|---|---|---|
| 1 | 9 | 1 | 7 | 3 | 4 | 4 | 6 | 2 | 8 |
| 7 | 3 | 9 | 1 | 7 | 6 | 4 | 4 | 6 | 7 |
| 3 | 7 | 5 | 9 | 1 | 2 | 6 | 4 | 5 | 3 |
| 5 | 3 | 7 | 7 | 9 | 8 | 2 | 3 | 5 | 5 |
| 4 | 4 | 2 | 8 | 1 | 9 | 1 | 6 | 6 | 4 |
| 6 | 2 | 8 | 3 | 7 | 8 | 9 | 1 | 4 | 6 |
| 2 | 8 | 5 | 5 | 3 | 2 | 8 | 9 | 1 | 2 |
| 8 | 7 | 3 | 5 | 6 | 5 | 2 | 8 | 9 | 1 |
| 9 | 1 | 7 | 6 | 4 | 5 | 3 | 2 | 8 | 9 |

The columns, rows and diagonals add to 495 ==> 18 ==> 9
The sum equals (99*100)/2 = 4950 ==> 18 ==> 9
The vertices add to 198 ==> 18 ==> 9
The center square equals 198 ==> 18 ==> 9
The sum minus the center square = 4752 ==> 18 ==> 9
The diagonals increase by 9 and 11
The vertices are male and female

| SDQ | NUMBER OF OCCURRENCES |
|---|---|
| 0 | 1 |
| 1 | 11 |
| 2 | 11 |
| 3 | 11 |
| 4 | 11 |
| 5 | 11 |
| 6 | 11 |
| 7 | 11 |
| 8 | 11 |
| 9 | 11 |

SUMMARY OF ODD AND EVEN ORDER LOST MAGIC SQUARES

| n | SUM | SDQ SUM | SUM/4 | SDQ SUM/4 | S | SDQ S | CENTER,SDQ | DIAGONALS |
|---|-----|---------|-------|-----------|---|-------|------------|-----------|
| 3 | 36 | 9 | 9 | 9 | 12 | 3 | 4 | 1,3 |
| 4 | 120 | 3 | 30 | 3 | 30 | 3 | 30=3 | 3,5 |
| 5 | 300 | 3 | 75 | 3 | 60 | 6 | 12=3 | 1,5 |
| 6 | 630 | 9 | 157.5 | 9 | 105 | 6 | 70=7 | 5,7 |
| 7 | 1176 | 6 | 294 | 6 | 168 | 6 | 24=6 | 1,7 |
| 8 | 2016 | 9 | 504 | 9 | 252 | 9 | 126=9 | 7,9 |
| 9 | 3240 | 9 | 810 | 9 | 360 | 9 | 40=4 | 1,9 |
| 10 | 4950 | 9 | 1237.5 | 9 | 495 | 9 | 198=9 | 9,11 |
| 11 | 7260 | 6 | 1815 | 6 | 660 | 3 | 60=6 | 1,11 |
| 12 | 10296 | 9 | 2574 | 9 | 858 | 3 |  | 11,13 |
| 13 | 14196 | 3 | 3549 | 3 | 1092 | 3 | 84=3 | 1,13 |
| 14 | 19110 | 3 | 4777.5 | 3 | 1365 | 6 |  | 13,15 |
| 15 | 25200 | 9 | 6300 | 9 | 1680 | 6 | 112=4 | 1,15 |
| 16 | 32640 | 6 | 8160 | 6 | 2040 | 6 |  | 15,17 |
| 17 | 41616 | 9 | 10404 | 9 | 2448 | 9 | 144=9 | 1,17 |
| 18 | 52326 | 9 | 13081.5 | 9 | 2907 | 9 |  | 17,19 |
| 19 | 64980 | 9 | 16245 | 9 | 3420 | 9 | 180=9 | 1,19 |
| 20 | 79800 | 6 | 19950 | 6 | 3990 | 3 |  | 19,21 |
| 21 | 97020 | 9 | 24255 | 9 | 4620 | 3 | 220=4 | 1,21 |

# Appendix 4

PROPERTIES OF ODD AND EVEN ORDER STANDARD MAGIC SQUARES

| n | SUM (St) | SDQ SUM | SUM/4 | SDQ (SUM/4) | S | SDQ(S) | CENTER,SDQ | DIAGONALS |
|---|---|---|---|---|---|---|---|---|
| 3 | 45 | 9 | 11.25 | 9 | 15 | 6 | 5 | 1,3 |
| 4 | 136 | 1 | 34 | 7 | 34 | 7 | 30=3 | 3,5 |
| 5 | 325 | 1 | 81.25 | 7 | 65 | 2 | 13=4 | 1,5 |
| 6 | 666 | 9 | 166.5 | 9 | 111 | 3 | 70=7 | 5,7 |
| 7 | 1225 | 1 | 306.25 | 7 | 175 | 4 | 25=7 | 1,7 |
| 8 | 2080 | 1 | 520 | 7 | 260 | 8 | 126=9 | 7,9 |
| 9 | 3321 | 9 | 830.25 | 9 | 369 | 9 | 41=5 | 1,9 |
| 10 | 5050 | 1 | 1262.5 | 7 | 505 | 1 | 198=9 | 9,11 |
| 11 | 7381 | 1 | 1845.25 | 7 | 671 | 5 | 61=7 | 1,11 |
| 12 | 10440 | 9 | 2610 | 9 | 870 | 6 | | 11,13 |
| 13 | 14365 | 1 | 3591.25 | 7 | 1105 | 7 | 85=4 | 1,13 |
| 14 | 19306 | 1 | 4826.5 | 7 | 1379 | 2 | | 13,15 |
| 15 | 25425 | 9 | 6356.25 | 9 | 1695 | 3 | 113=5 | 1,15 |
| 16 | 32896 | 1 | 8224 | 7 | 2056 | 4 | | 15,17 |
| 17 | 41905 | 1 | 10476.25 | 7 | 2465 | 8 | 145=1 | 1,17 |
| 18 | 52650 | 9 | 13162.5 | 9 | 2925 | 9 | | 17,19 |
| 19 | 65341 | 1 | 16335.25 | 7 | 3439 | 1 | 181=1 | 1,19 |
| 20 | 80200 | 1 | 20050 | 7 | 4010 | 5 | | 1921 |
| 21 | 97461 | 9 | 24365.25 | 9 | 4641 | 6 | 221=5 | 1,21 |
| 22 | | 1 | | | | 7 | | |
| 23 | | 1 | | | | 2 | | |
| 24 | | 9 | | | | 3 | | |
| 25 | | 1 | | | | 4 | | |
| 26 | | 1 | | | | 8 | | |

| | | | | | | | | |
|---|---|---|---|---|---|---|---|---|
| 27 | | 9 | | | | 9 | | |
| 28 | | 1 | | | | 1 | | |
| 29 | | 1 | | | | 5 | | |
| 30 | | 9 | | | | 6 | | |
| 31 | | 1 | | | | 7 | | |
| 32 | | 1 | | | | 2 | | |

## SUMMARY OF ODD AND EVEN ORDER LOST MAGIC SQUARES

| n | SUM | SDQ SUM | SUM/4 | SDQ SUM/4 | S | SDQ S | CENTER,SDQ | DIAGONALS |
|---|---|---|---|---|---|---|---|---|
| 3 | 36 | 9 | 9 | 9 | 12 | 3 | 4 | 1,3 |
| 4 | 120 | 3 | 30 | 3 | 30 | 3 | 30=3 | 3,5 |
| 5 | 300 | 3 | 75 | 3 | 60 | 6 | 12=3 | 1,5 |
| 6 | 630 | 9 | 157.5 | 9 | 105 | 6 | 70=7 | 5,7 |
| 7 | 1176 | 6 | 294 | 6 | 168 | 6 | 24=6 | 1,7 |
| 8 | 2016 | 9 | 504 | 9 | 252 | 9 | 126=9 | 7,9 |
| 9 | 3240 | 9 | 810 | 9 | 360 | 9 | 40=4 | 1,9 |
| 10 | 4950 | 9 | 1237.5 | 9 | 495 | 9 | 198=9 | 9,11 |
| 11 | 7260 | 6 | 1815 | 6 | 660 | 3 | 60=6 | 1,11 |
| 12 | 10296 | 9 | 2574 | 9 | 858 | 3 | | 11,13 |
| 13 | 14196 | 3 | 3549 | 3 | 1092 | 3 | 84=3 | 1,13 |
| 14 | 19110 | 3 | 4777.5 | 3 | 1365 | 6 | | 13,15 |
| 15 | 25200 | 9 | 6300 | 9 | 1680 | 6 | 112=4 | 1,15 |
| 16 | 32640 | 6 | 8160 | 6 | 2040 | 6 | | 15,17 |
| 17 | 41616 | 9 | 10404 | 9 | 2448 | 9 | 144=9 | 1,17 |
| 18 | 52326 | 9 | 13081.5 | 9 | 2907 | 9 | | 17,19 |

| 19 | 64980 | 9 | 16245 | 9 | 3420 | 9 | 180=9 | 1,19 |
| 20 | 79800 | 6 | 19950 | 6 | 3990 | 3 | | 19,21 |

## NUMBER PATTERNS

| n | 3 | 4 | 5 | 6 | 7 | 8 | 9 | 10 | 11 | 12 | 13 | 14 | 15 | 16 |
|---|---|---|---|---|---|---|---|----|----|----|----|----|----|----|
| n SDQ | 3 | 4 | 5 | 6 | 7 | 8 | 9 | 1 | 2 | 3 | 4 | 5 | 6 | 7 |

**S T A N D A R D   M A G I C   S Q U A R E S**

| SUM SDQ | 9 | 1 | 1 | 9 | 1 | 1 | 9 | 1 | 1 | 9 | 1 | 1 | 9 | 1 |
|---------|---|---|---|---|---|---|---|---|---|---|---|---|---|---|
| S SDQ | 6 | 7 | 2 | 3 | 4 | 8 | 9 | 1 | 5 | 6 | 7 | 2 | 3 | 4 |

**L O S T   M A G I C   S Q U A R E S**

| SUM SDQ | 9 | 3 | 3 | 9 | 6 | 9 | 9 | 9 | 6 | 9 | 3 | 3 | 9 | 6 |
|---------|---|---|---|---|---|---|---|---|---|---|---|---|---|---|
| S SDQ | 3 | 3 | 6 | 6 | 6 | 9 | 9 | 9 | 3 | 3 | 3 | 6 | 6 | 6 |

# MAGIC SQUARES: POSITIVE AND NEGATIVE NUMBERS

### ODD ORDER: 3x3 PLUS-MINUS MAGIC SQUARE

One of the attributes of regular magic squares is that they use consecutive numbers. Suppose we now consider the use of both positive and negative consecutive numbers. Consider the 3x3 odd order standard magic square:

| 4 | 9 | 2 |
|---|---|---|
| 3 | 5 | 7 |
| 8 | 1 | 6 |

If we add -5 to the number in each cell, we get:

| -1 | 4  | - 3 |
|----|----|-----|
| -2 | 0  | 2   |
| 3  | -4 | 1   |

This is a very unusual magic square since all diagonals and columns and rows, including the sum total, all add to 0. The absolute values of the maximum cell numbers are the same, 4 and -4, are above and below the central 0, respectively. The positive central cross adds to 6.

Notice that one of the cells has a zero, not unlike the 0 of the lost magic squares. In fact, starting with a 3x3 lost magic square;

| 3 | 8 | 1 |
|---|---|---|
| 2 | 4 | 6 |
| 7 | 0 | 5 |

we need to add -4 to the numbers in each cell to produce the same 0 total magic square;

| -1 | 4  | -3 |
|----|----|----|
| -2 | 0  | 2  |
| 3  | -4 | 1  |

If we require that we have consecutive positive and negative numbers, we must include the zero. In this respect we will always have a zero, as in the lost magic squares.

## EVEN ORDER: 4X4 PLUS AND MINUS MAGIC SQUARE
Consider now a 4x4 lost magic square (the same analysis holds for a standard 4x4 magic square):

| 3  | 13 | 14 | 0  |
|----|----|----|----|
| 8  | 6  | 5  | 11 |
| 4  | 10 | 9  | 7  |
| 15 | 1  | 2  | 12 |

Since there are an even number of cells, there cannot both be a 0 and the same absolute magnitude for the positive and negative numbers. If we omit the 0 by going from -1 to 1 in the sequence of counting, then we could have the same absolute magnitudes, thus;

| -5 | 6 | 7 | -8 |
|----|----|----|----|
| 1 | -2 | -3 | 4 |
| -4 | 3 | 2 | -1 |
| 8 | -7 | -6 | 5 |

We see that the price for symmetry of magnitude is to lose the consecutive numbering, and thus we have lost the 0. The rows, columns, diagonals and total sum add to 0.

For the even order magic squares using consecutive negative and positive numbering, we cannot start and end at the same magnitude, for example, here, in the 4x4, we could go from -8 to +7, or from -7 to +8. We also recognize, that since we are not using a symmetric negative and positive numbering, the sums can not add to 0. Choosing the range from -7 to +8 we get:

| -4 | 6 | 7 | -7 |
|----|----|----|----|
| 1 | -1 | -2 | 4 |
| -3 | 3 | 2 | 0 |
| 8 | -6 | -5 | 5 |

We note that the sum of the rows, columns, and diagonals add to 2, and the total sum is 8.

## ODD ORDER: 5X5 PLUS-MINUS MAGIC SQUARE

| 10 | 23 | 6 | 19 | 2 |
|----|----|----|----|----|
| 3 | 11 | 24 | 7 | 15 |
| 16 | 4 | 12 | 20 | 8 |
| 9 | 17 | 0 | 13 | 21 |
| 22 | 5 | 18 | 1 | 14 |

In this lost magic square, (the same principle holds for the 5x5 standard magic square), we need only add -12 to each cell, to achieve the desired plus-minus magic square:

| -2 | 11 | -6 | 7 | -10 |
|----|----|----|----|-----|
| -9 | -1 | 12 | -5 | 3 |
| 4 | -8 | 0 | 8 | -4 |
| -3 | 5 | -12 | 1 | 9 |
| 10 | -7 | 6 | -11 | 2 |

As with the 3x3 plus-minus square, all rows, columns, diagonals and total sum add to 0, and the numbers run consecutively. Note the symmetry of the numbers in the cells equidistant from the center, 0. Rows and columns equidistant form the center alternate numbers and signs. Thus, the upper row and the bottom row, each have the numbers 2, 11, 6, 7, and 10, in reverse order and with opposite signs. The max and min numbers, plus and minus 12 are above and below the central 0, respectively. The positive numbers of the central cross add to +20 and -20, respectively.

As a general rule, any odd order magic square, may be converted to a plus-minus magic square by adding the appropriate negative number to each cell. We then end up with a square that has a zero in one of its cells, and whose columns, rows, diagonals and total sum are 0.

ODD ORDER: 5X5 PLUS-MINUS MAGIC SQUARE: SDQ

| -2 | 2 | -6 | 7 | -1 |
|----|----|----|----|-----|
| -9 | -1 | 3 | -5 | 3 |
| 4 | -8 | 0 | 8 | -4 |
| -3 | 5 | -3 | 1 | 9 |
| 1 | -7 | 6 | -2 | 2 |

ODD ORDER: 7X7 LOST MAGIC SQUARE

| 21 | 46 | 15 | 40 | 9  | 34 | 3  |
|----|----|----|----|----|----|----|
| 4  | 22 | 47 | 16 | 41 | 10 | 28 |
| 29 | 5  | 23 | 48 | 17 | 35 | 11 |
| 12 | 30 | 6  | 24 | 42 | 18 | 36 |
| 37 | 13 | 31 | 0  | 25 | 43 | 19 |
| 20 | 38 | 7  | 32 | 1  | 26 | 44 |
| 45 | 14 | 39 | 8  | 33 | 2  | 27 |

For this 7x7 lost magic square, we simply add -24 to each cell to produce the plus-minus magic square.

ODD ORDER: 7X7 PLUS-MINUS MAGIC SQUARE

| -3  | 22  | -9  | 16  | -15 | 10  | -21 |
|-----|-----|-----|-----|-----|-----|-----|
| -20 | -2  | 23  | -8  | 17  | -14 | 4   |
| 5   | -19 | -1  | 24  | -7  | 11  | -13 |
| -12 | 6   | -18 | 0   | 18  | -6  | 12  |
| 13  | -11 | 7   | -24 | 1   | 19  | -5  |
| -4  | 14  | -17 | 8   | -23 | 2   | 20  |
| 21  | -10 | 15  | -16 | 9   | -22 | 3   |

Again, we note the incredible symmetry of this magic square, with all rows, columns, diagonals and total sum adding to 0. The max and min numbers, 24 and -24, are above and below the central 0, respectively. The central cross positive numbers add to 42.

## ODD ORDER: 7X7 PLUS-MINUS MAGIC SQUARE; SDQ

| -3 | 4  | -9 | 7  | -6 | 1  | -3 |
|----|----|----|----|----|----|----|
| -2 | -2 | 5  | -8 | 8  | -5 | 4  |
| 5  | -1 | -1 | 6  | -7 | 2  | -4 |
| -3 | 6  | -9 | 0  | 9  | -6 | 3  |
| 4  | -2 | 7  | -6 | 1  | 1  | -5 |
| -4 | 5  | -8 | 8  | -5 | 2  | 2  |
| 3  | -1 | 6  | -7 | 9  | -4 | 3  |

## ODD ORDER: 9X9 STANDARD MAGIC SQUARE

| 37 | 78 | 29 | 70 | 21 | 62 | 13 | 54 | 5  |
|----|----|----|----|----|----|----|----|----|
| 6  | 38 | 79 | 30 | 71 | 22 | 63 | 14 | 46 |
| 47 | 7  | 39 | 80 | 31 | 72 | 23 | 55 | 15 |
| 16 | 48 | 8  | 40 | 81 | 32 | 64 | 24 | 56 |
| 57 | 17 | 49 | 9  | 41 | 73 | 33 | 65 | 25 |
| 26 | 58 | 18 | 50 | 1  | 42 | 74 | 34 | 66 |
| 67 | 27 | 59 | 10 | 51 | 2  | 43 | 75 | 35 |
| 36 | 68 | 19 | 60 | 11 | 52 | 3  | 44 | 76 |
| 77 | 28 | 69 | 20 | 61 | 12 | 53 | 4  | 45 |

## ODD ORDER: 9x9 PLUS MINUS MAGIC SQUARE
(add -41 to each cell of the standard 9x9)

| -4  | 37  | -12 | 29  | -20 | 21  | -28 | 13  | -36 |
|-----|-----|-----|-----|-----|-----|-----|-----|-----|
| -35 | -3  | 38  | -11 | 30  | -19 | 22  | -27 | 5   |
| 6   | -34 | -2  | 39  | -10 | 31  | -18 | 14  | 26  |
| -25 | 7   | -33 | -1  | 40  | -9  | 23  | -17 | 12  |
| 16  | -24 | 8   | -32 | 0   | 32  | -8  | 24  | -16 |
| -15 | 17  | -23 | 9   | -40 | 1   | 44  | -7  | 25  |
| 26  | -14 | 18  | -31 | 10  | -39 | 2   | 34  | -6  |
| -5  | 27  | -22 | 19  | -30 | 11  | -38 | 3   | 35  |
| 36  | -13 | 28  | -21 | 20  | -29 | 12  | -37 | 4   |

The incredible symmetry makes it easy to spot any typographical error in the table. All rows, columns, diagonals, and total sum add to 0. The max and min numbers, plus and minus 40 are above and below the central 0. The center square adds to 72=>9.

## MAGIC SQUARES AND THE SOLAR SYSTEM
The following is an ancient association of the standard magic squares and members of the solar system.

| n  | S   | SDQ(S) | BODY      | PLUTARCH | SDQ(PLUT) |
|----|-----|--------|-----------|----------|-----------|
| 10 | 505 | 1      | EARTH     | 9        | 9         |
| 5  | 65  | 2      | MARS      |          |           |
| 6  | 111 | 3      | SUN       | 729      | 9         |
| 7  | 175 | 4      | VENUS     | 243      | 9         |
| 11 | 671 | 5      |           |          |           |
| 3  | 15  | 6      | SATURN    |          |           |
| 4  | 34  | 7      | JUPITER   |          |           |
| 8  | 260 | 8      | MERCURY   | 81       | 9         |
| 9  | 369 | 9      | MOON      | 27       | 9         |
| 12 | 870 | 6      | ANTICHTHON| 3        | 3         |

Let us examine this table by referring to the results of standard or regular magic squares.

Regular Squares                    Lost Square

| S | n | SDQ(S) | S | n | SDQ(S) |
|---|---|--------|---|---|--------|
| 15 | 3 | 6 | 12 | 3 | 3 |
| 34 | 4 | 7 | 30 | 4 | 3 |
| 65 | 5 | 2 | 60 | 5 | 6 |
| 111 | 6 | 3 | 105 | 6 | 6 |
| 175 | 7 | 4 | 168 | 7 | 6 |
| 260 | 8 | 8 | 252 | 8 | 0 |
| 369 | 9 | 0 | 360 | 9 | 0 |
| 505 | 10 | 1 | 495 | 10 | 0 |
| 671 | 11 | 5 | 660 | 11 | 3 |
| | | | | | |
| 870 | 12 | 6 | 858 | 12 | 3 |
| 1105 | 13 | 7 | 1092 | 13 | 3 |
| 1379 | 14 | 2 | 1365 | 14 | 6 |
| 1695 | 15 | 3 | 1680 | 15 | 6 |
| 2056 | 16 | 4 | 2040 | 16 | 6 |
| 2465 | 17 | 8 | 2448 | 17 | 0 |
| 2925 | 18 | 0 | 2907 | 18 | 0 |
| 3439 | 19 | 1 | 3420 | 19 | 0 |
| 4010 | 20 | 5 | 3990 | 20 | 3 |
| | | | | | |
| 4641 | 21 | 6 | 4620 | 21 | 3 |
| 5335 | 22 | 7 | 5313 | 22 | 3 |

This table shows results for the regular (standard) and the lost magic squares. The first column shows S, the magic constant-the sum of the sides for each order number, n, given in next column. The third column is the SDQ of S.

We notice that the SDQ of S includes all of the digits for the order numbers n= 3, ..., 11. This pattern of SDQ(S) repeats with n=12. We can see that by taking the SDQ(n), that the pattern again repeats, n=12=>3, and this is exactly the value for SDQ(S) for n=3, namely, 6. I believe that the ancients used this ordering of the SDQ(S), for their order of the planets, as they knew them.

| SDQ(S) | BODY |
|:------:|:----:|
| 1 | Earth |
| 2 | Mars |
| 3 | Sun |
| 4 | Venus |
| 5 |  |
| 6 | Saturn |
| 7 | Jupiter |
| 8 | Mercury<br>MERCURY |
| 9 | Moon |
| 6 | Antichthon |

The late Aryeh Kaplan, goes further and associates additional magic squares with the Sefirot. More on that in the chapter on the Tree of Life.

The association of the order of the square with a particular body is ancient, but we see that by looking at the SDQ of the sum, S, we have another way of ordering the bodies. Two interesting aspects of the SDQ ordering are; a missing body for the SDQ of 5, and a double set of bodies, Saturn and Antichthon, for the SDQ of 6. Antichthon is said to be a planet that is diametrically opposite to earth.

It should be immediately obvious that the outer planets, Uranus, Neptune and Pluto are not included in this correspondence. A possible correspondence with the SDQ of 5, is the body called Marduk, by the Sumerians. More on that topic in a forthcoming book.

# The Fibonacci Square

The Fibonacci series is obtained starting with any two numbers and then adding them to produce the next number. Starting with 0 and then 1, we get the following:

0; 1 1 2 3 5 8 13 21 34 55 89 144 233 377 610 987 1,597 2,584 4,181 6,765 10,946 17,711 28,657 46,368; 75,025 121,393 196,418

Using SDQ, this series becomes: 0; 1 1 2 3 5 8 4 3 7 1 8 9 8 8 7 6 4 1 5 6 2 8 1 9; 1 1 2, and we note that the series repeats again with the double one's.

Thus, the Fibonnaci series, upon which the Golden Proportion rests, is a repetitive series, of period 24, when reduced through SDQ. Note that the period of 24 itself reduces to an SDQ of 6.

Consider the standard 3x3 magic square:

| 4 | 9 | 2 |
|---|---|---|
| 3 | 5 | 7 |
| 8 | 1 | 6 |

where all of the rows, columns and diagonals add to the magic sum, 15.

As there are nine terms, let us consider a Fibonacci series of the Golden Proportion, $\phi$, where $\phi = 1.62$. We will use a total of nine terms, four each that are symmetric about the central term 1, namely: $\phi^{-4}$, $\phi^{-3}$, $\phi^{-2}$, $\phi^{-1}$, $\phi^0$, $\phi^1$, $\phi^2$, $\phi^3$, $\phi^4$. ( Note that $\phi^0 = 1$). That this is a Fibonacci series can be seen by noting that each term is the sum of the two preceding terms, $\phi^{-4} + \phi^{-3} = \phi^{-2}$, and so on. If we order these nine terms starting with $\phi^{-4}$ as 1, proceeding to $\phi4$ as 9, we can replace the numbers in the 3x3 square with their $\phi$ equivalents. The magic square then becomes:

| $\phi^{-1}$ | $\phi^4$ | $\phi^{-3}$ |
|---|---|---|
| $\phi^{-2}$ | $\phi^0$ | $\phi^2$ |
| $\phi^3$ | $\phi^{-4}$ | $\phi^1$ |

This magic square has the remarkable properties that every row, column and diagonal has a product of 1. We shall call this product, the magic product number. Also, all terms symmetric about the center, have a product of 1. In fact the product of all the contents of the square is 1.

In this manner we can build any odd order Fibonacci magic square. A n = 5 order square, having 25 terms, would need 12 terms of negative $\phi$ exponentiation and twelve terms of positive $\phi$ exponentiation, together with 1 or φ0. As in n = 3, the n = 5 square has magic product constant = 1. ( The overall product is again 1).

Replacing φ by any other base, such as 2, the magic product will still be 1.

| $2^{-1}$ | $2^4$ | $2^{-3}$ |
|---|---|---|
| $2^{-2}$ | $2^0$ | $2^2$ |
| $2^3$ | $2^{-4}$ | $2^1$ |

The base 2, however, is not a Fibonacci series, since 20 + 21 does not equal 22. Only the φ series is a Fibonacci series.

$729 = 9 \times 9 \times 9 = 3 \times 3 \times 3 \times 3 \times 3 \times 3 = 3^6$.

# PROPERTIES OF SQUARE NUMBERS

All square numbers have additional properties. Starting with 1,2,3, ... their squares are: 1, 4, 9, 16, 25, 36, 49, 64, 81, 100, 121, 144, 169, 196, 225, 256, 289, 324, 361, 400, 441, 484, 529, 576, 625, 676, 729, 784, 841, 900, 961, 1024, etc. Their corresponding SDQ are: 1, 4, 9, 7, 7, 9, 4, 1, 9,  1, 4, 9, 7, 7, 9, 4, 1, 9,  1, 4, 9, 7, 7, 9, 4, 1, 9,  1, 4, 9, 7, 7, ...

We have left a space where the 9 term pattern repeats itself. Thus, the period is 9, the pattern is 1, 4, 9, 7, 7, 9, 4, 1, 9, and these numbers add to 51 with an SDQ(6).

Only the digits 1, 4, 7, and 9 are used. 1+4+7+9=21=3.

The roots of the squares, following the pattern are: 1, 2, 3, 4, 5, 6, 7, 8, 9,  10, 11, 12, 13, 14, 15, 16, 17, 18,    19, 20, 21, 22, 23, 24, 25, 26, 27,   ... The SDQ of the patterns of the roots are:  45 = 9, 126 = 9, 207 = 9, all 9's.

# PROPERTIES OF MAGIC SQUARE NUMBERS

Magic squares start with the square number 9 (3x3), so that now, the square numbers follow the sequence; 9, 16, 25, 36, 49, 64, 81, 100, 121,  144, 169, 196, 225, 256, 289, 324, 361, 400,  441, 484, 529, 576, 625, 676, 729, 784, 841,  900, 961, 1024, etc.

SDQ= 9, 7, 7, 9, 4, 1, 9, 1, 4,  9, 7, 7, 9, 4, 1, 9, 1, 4,
9, 7, 7, 9, 4, 1, 9, 1, 4,  9, 7, 7,

Once again the period is 9, the pattern being 9, 7, 7, 9, 4, 1, 9, 1, 4, and theses numbers add to 501 with an SDQ also of (6).

Only the digits 1, 4, 7, and 9 are used. 1+4+7+9=21=3.

The roots of the squares, following the pattern are: 3, 4, 5, 6, 7, 8, 9, 10, 11, 12, 13, 14, 15, 16, 17, 18, 19, 20,    21, 22, 23, 24, 25, 26, 27, 28, 29, ... The SDQ of the patterns of the roots are: 63 = 9, 144 => 9, 225 => 9, all 9's.

Every third root, has an SDQ of 9, and the SDQ of its square is also 9, which makes sense, since any SDQ of 9, when squared still yields an SDQ of 9.
Look at the SDQ for the squares, see paper.
Each square then has overall properties of one of four powers, namely; 1, 4, 7, or 9. 1 3 6 1 6 3 1 9 9
There are a number of creations, those by the Elohim and those by YHVH Elohim. Look at the creations(?) by YHVH.
Analyze all creations by their creators.
Who, how, when, where, why, etc. Ask all the questions.

Any number that equilibrates two others is special. a acting on b produces the effect c.
3  8  1
2  4  6
7  0  5    S=12        sets the pattern. Can also increase each element by + or - n.
Thus also
4  9  2                -1  4  -3
3  5  7                -2  0  2
8  1  6    S=15         3  -4  1    S=0. All adds to 0.
Thus any number may be in the center.
2  7  0
1  3  5
6  -1  4    S=9
1  6  -1
0  2  4
5  -2  3    S=6.   S always equals 3x center value, thus always a multiple of 3.
0  5  -2
-1  1  3
4  -3  2    S=3
17  22  15
16  18  20
21  14  19  S=54

## THIRD ORDER PRIME NUMBER

| 67 | 1 | 43 |
|----|----|----|
| 13 | 37 | 61 |
| 31 | 73 | 7 |

S = 111

C = 37

St = 333

extremes = 74

center balances the extremes

## THIRD ORDER PRIME NUMBER: SDQ

SDQ

| 4 | 1 | 7 |
|---|---|---|
| 4 | 1 | 7 |
| 4 | 1 | 7 |

SDQ S = 3; C = 1; SDQ St = 36 = 9; SDQ extremes = 2.

square value: $2(n2-1)$

singly-even magic squares = divisible by 2, not 4

6*6, 10*10, 14*14, 18*18, etc

3 0 3, 5 0 5, (5 0ver and against 5), 7 0 7, m f m

doubly-even magic squares = divisible by 4

4*4, 8*8, 12*12, 16*16, 20*20, 24*24, 28*28, 32*32, etc

2 0 2, 4 0 4, 6 0 6, f f f, etc

even single are thus m f m

even double are thus f f f

semimagic squares = row and columns ok, one or more diagonals wrong.

mgc sqrs

## PROPERTIES OF SQUARE NUMBERS

All square numbers have additional properties. Starting with 1,2,3, ... their squares are: 1, 4, 9, 16, 25, 36, 49, 64, 81, 100, 121, 144, 169, 196, 225, 256, 289, 324, 361, 400, 441, 484, 529, 576, 625, 676, 729, 784, 841, 900, 961, 1024, etc. Their corresponding SDQ are: 1, 4, 9, 7, 7, 9, 4, 1, 9,   1, 4, 9, 7, 7, 9, 4, 1, 9,   1, 4, 9, 7, 7, 9, 4, 1, 9,   1, 4, 9, 7, 7, ...

We have left a space where the 9 term pattern repeats itself. Thus, the period is 9, the pattern is 1, 4, 9, 7, 7, 9, 4, 1, 9, and these numbers add to 51 with an SDQ(6).

Only the digits 1, 4, 7, and 9 are used. 1+4+7+9=21=3.

The roots of the squares, following the pattern are: 1, 2, 3, 4, 5, 6, 7, 8, 9,  10, 11, 12, 13, 14, 15, 16, 17, 18,    19, 20, 21, 22, 23, 24, 25, 26, 27,   ... The SDQ of the patterns of the roots are:  45 = 9, 126 = 9, 207 = 9, all 9's.

## PROPERTIES OF MAGIC SQUARE NUMBERS

Magic squares start with the square number 9 (3x3), so that now, the square numbers follow the sequence; 9, 16, 25, 36, 49, 64, 81, 100, 121,  144, 169, 196, 225, 256, 289, 324, 361, 400,  441, 484, 529, 576, 625, 676, 729, 784, 841,  900, 961, 1024, etc.

SDQ= 9, 7, 7, 9, 4, 1, 9, 1, 4; 9, 7, 7, 9, 4, 1, 9, 1, 4; 9, 7, 7, 9, 4, 1, 9, 1, 4;  9, 7, 7,

Once again the period is 9, the pattern being 9, 7, 7, 9, 4, 1, 9, 1, 4, and theses numbers add to 501 with an SDQ also of (6).

Only the digits 1, 4, 7, and 9 are used. 1+4+7+9=21=3.

The roots of the squares, following the pattern are: 3, 4, 5, 6, 7, 8, 9, 10, 11,  12, 13, 14, 15, 16, 17, 18, 19, 20,    21, 22, 23, 24, 25, 26, 27, 28, 29, ... The SDQ of the patterns of the roots are: 63 = 9, 144 => 9, 225 => 9, all 9's.

Every third root, has an SDQ of 9, and the SDQ of its square is also 9, which makes sense, since any SDQ of 9, when squared still yields an SDQ of 9.

## THIRD ORDER PRIME NUMBER

| 67 | 1 | 43 |
|----|----|----|
| 13 | 37 | 61 |
| 31 | 73 | 7 |

S = 111

C = 37

St = 333

extremes = 74

center balances the extremes

## THIRD ORDER PRIME NUMBER: SDQ

SDQ

| 4 | 1 | 7 |
|----|----|----|
| 4 | 1 | 7 |
| 4 | 1 | 7 |

SDQ S = 3; C = 1; SDQ St = 36 = 9; SDQ extremes = 2.

square value: 2(n2-1)

singly-even magic squares = divisible by 2, not 4

6*6, 10*10, 14*14, 18*18, etc

3 0 3, 5 0 5, (5 0ver and against 5), 7 0 7, m f m

doubly-even magic squares = divisible by 4

4*4, 8*8, 12*12, 16*16, 20*20, 24*24, 28*28, 32*32, etc

2 0 2, 4 0 4, 6 0 6, f f f, etc

even single are thus m f m

even double are thus f f f

semimagic squares = row and columns ok, one or more diagonals wrong.

mgc sqrs

## DEFINITIONS

normal = formed of first $n^2$ natural numbers

magic constant = sum of row or column or diagonal

$n^2$ = nth order

complement = the number that you must add to make the numbers add to $n^2 +1$. Starting at 1. Starting at 0, must add to $n^2 -1$

Rem: [add most general sum, $S = nA + B(n/2)(n^2 - 1)$, where

A = initial number; B = increment; n = order]

If there were a 2x2 magic square, its sum (A=0) would be 3, where B=1.

Look at SDQ of 2S when A=0.

[FAR] Add section on cell size and relativity characteristics]

REM: See if the squares can be rearranged to conform to the 9x9 pattern when reduced to a SDQ!

The multiples of 4 are DOUBLE OCTAVES!!

0, 12, 20, are three mothers, all evenly-even magic squares.

tet=8 is a simple, and peh=16 is a double.

What about squares composed of factorials?

## ERRATA

$S = nA + B(n/2)(n^2 - 1)$, where

A = initial number; B = increment; n = order

or $S = n[A + (B/2)(n^2 - 1)]$

If there were a 2x2 magic square, its lost sum (A=0) would be 3, where B=1. And its standard sum (A=1) would be 5, B=1.

Standard: n=3, A=1, B=1; S= 3 + (3/2)(8) = 15. Correct!

Standard: n=3, A=1, B=2; S= 3 +2*(3/2)*8 = 27.

Standard: n=3, A=1, B=4; S= 3 +4*(3/2)*8 = 51.

Lost: n=3, A=0, B=1; S=(3/2)(8) = 12. Correct!

Lost: n=3, A=0, B=2; S=2*(3/2)*8 = 24.

Lost: n=3, A=0, B=4; S=4*(3/2)*8 = 48.

n=3:

S = 3A + B(3/2)*8 = 3A +12B = 3(A + 4B)

S(standard) = 3(1+4B)

S(lost) = 12B

Suppose we wish to examine a B=1, and S = 0. Then, 3(A+4)=0, or

3A + (3/2)*8 = 0, or 3A + 12 = 0, with the solution that A = -4.

n=4:

S = 4A + B(4/2)(16 - 1) = 4A + 2B(15) = 2(2A + 15B)

Then A = -7.5B.

Suppose we wish to examine a B=1, and S = 0. Then

2A + 15 = 0, or A = -7.5

-7.5,-6.5,-5.5,-4.5,-3.5,-2.5,-1.5,-.5, +.5, +1.5, +2.5, +3.5, +4.5, +5.5, +6.5, +7.5.

If B=2, then A= -15.

-15, -13, -11, -9, -7, -5, -3, -1, 1, 3, 5, 7, 9, 11, 13, 15

[Looks like a quantum number projection along the z-axis]

[end rem:]

Look at the SDQ for the squares, see paper.

Each square then has overall properties of one of four powers, namely; 1, 4, 7, or

9. 1 3 6 1 6 3 1 9 9

Any number that equilibrates two others is special. a acting on b produces the effect c.

3  8  1

2  4  6

7  0  5        S=12          sets the pattern. Can also increase each element by + or - n.

Thus also

4  9  2                  -1  4  -3

3  5  7                  -2  0   2

8  1  6        S=15        3  -4   1    S=0. All adds to 0.

Thus any number may be in the center.

2  7  0
1  3  5
6 -1  4    S=9


1  6 -1
0  2  4
5 -2  3    S=6.   S always equals 3x center value, thus always a multiple of 3.


0  5 -2
-1 1  3
4 -3  2    S=3
17 22  15
16 18  20
21 14  19  S=5

# Appendix 5

## CHINESE MAGIC SQUARES

### THE I CHING

The I Ching or Book of Changes is credited to a legendary Emperor of China, Fu Hsi, c -3000. Fu Hsi is also credited with forming the first civilized society within China. We find ourselves on this planet as a rotating, revolving, and living geometric system. Every aspect of this planet is governed by cycles, the obvious ones being time but there are also space cycles. These are repetitive patterns such as lines of latitude and longitude that mensurate all changes. The I Ching is considered to be the set of principles for behavior within this planetary society. We are analyzing it with the aim of determining its basic mathematical properties both quantitative and qualitative.

Its long history as a tool for prognostication carries little if any interest at the present writing. The world of psychic phenomena, of which prognostication is a part, is a major study in itself. At the present time insufficient data of scientific quality exists for even a suspicion of any correlations. However, at some future time, the entire subject of psychic phenomena is requiring of study using the principles and techniques of Divine Science.

The I Ching manifests from a duality while the T=ai Hsüan Ching, see next section, manifests from a trinity. It is from either of these that all creations stem, in fact it is the combination of the power of two with the power of three that is the basis for all music theory. The octave, or manifestation of duality produces no new tones but is considered as the womb for all generations. It is the three that fertilizes this duality and from which, following the principles of averaging in nature, produce all others. The duality relates to the female principle and the trinity to the male principle.

Each line can represent a different level of existence, and for the Chinese systems of I Ching and

T=ai Hsüan Ching, the bottom line, the starting line, represents the earthly influences, the next line above represents humanly influences, and the third line represents heavenly influences. These influences are like forces that act on a system. They are vectors, having both magnitude and direction, and have a resultant. However, there is an additional force of will that can alter their resultant vector.

This is precisely the realm of Divine Science, wherein it is a physical-like system, with physical-like forces, that includes the forces of will for humanly influence, and the force of destiny for heavenly influences. It is the humanly force of will that compels a motivated child to stand, walk, climb, and achieve. It is the divine heavenly force that acts on the system of sperm and egg and participates in the chemical and physiological development of the foetus.

## MATHEMATICAL SYSTEMS

Musical systems are based upon regular numbers, those numbers that can be formed from 2p3q5r. The largest of value for 2p3q5r is called the Tonal Index and limits the total number of tones.   The sexagesimal system is based on powers of the regular number 60, since 60 = 223151, (p=2, q=1, r=1).

As we have noted the ancient musical system is based on powers of 2 and 3, namely 2p and 3q. Numbers that differ by multiples or sub multiples of 2 are considered to be identical musically, as they are just different octaves. It is from either of these that all creations stem, in fact it is the combination of the power of two with the power of three that is the basis for all music theory. The octave, or manifestation of duality produces no new tones but is considered as the womb for all generations. It is the three that fertilizes this duality and from which, following the principles of averaging in nature, produce all others. The duality relates to the female principle and the trinity to the male principle. This is what the power of generation is based upon, the power and the base. It is all both musical, mathematical and apparent.

## MATHEMATICS AND THE I CHING

Let us start with the I Ching consisting of unity that divides into two parts, the female Yin, and the male Yang. The Yin is a line with a space, -----__  _, and the Yang is a solid line, ------. We can add an additional line above each, thus there will now be four possibilities, two have Yang at the bottom and two have Yin at the bottom. The bottom line is considered the starting line, and will be called line 1. We can let 1 represent a Yang, or solid line, and 2 a Yin, or two part line.

Two line systems are called bigrams. There are thus four bigrams, where we list the bottom line first, thus: 1 1; 1 2; 2 1; 2 2. These will appear as follows: ___   _

_        ___   _  _
       ___ ; ___ ; and       _ _;   _  _ . A single line has 21 = 2 different combinations, while two lines have 22 = 4 different combinations. 1 1 is called

Greater Yang, and 2 2 is called Greater Yin. 1 2 is called Lesser Yang, and 2 1 is called Lesser Yin.

We can proceed to add an additional line, and this brings the possible different combinations to 23 = 8. These are called Trigrams.

Adding a fourth line, we get 24 = 16 different combinations. These are called Tetragrams.

The pattern is simple, Yin and Yang are the base of 2, and the number of lines is the exponent. Five lines, pentagrams, would have 25 = 32 different combinations, and six lines, hexagrams, would have 64 different combinations. It is easy to see the relationship of this procedure with octave doubling. Any number greater than four, that has an integer square root, can be used in a magic square. Thus, the tetragram and the hexagram, 4 lines and 6 lines, can be set into a 4 x 4 and 8 x 8 magic squares.


## THE T=AI HSÜAN CHING

The T=ai Hsüan Ching is an ancient book of divination and philosophy and was assembled by Yang Hsiung around two thousand years ago. The text is divided into 81 sections known as Shou. These Shou are numbered by a system of four-lined diagrams, called tetragrams. The four lines are: the top or principal line indicating a square, direction or compass point; the second line indicating a division; the third line indicating a department; and the fourth and bottom line, indicating a family. Each of the lines can be complete, in two parts or in three parts. In the sequence, the first Shou is composed of four solid lines.

According to Yang Hsiung, there is an additional force, the Jen, that balances the forces of T=ien and Ti. It is significant that T=ien and Ti are not considered as positive and negative, just extreme forces. Thus, two weights that balance on a scale are both positive. The meaning of zero is therefore not a unique mathematical concept, as any two equal weights can balance to zero. Not all zeros are alike.


## MATHEMATICS AND THE T'AI HSÜAN CHING

The T'ai Hsüan Ching consists of a one, two, or three part line. Thus, ------, --- ---, and -- -- --. For bigrams, we have 9 possibilities,

-------- --- --- -- -- -- --------- --- --- -- -- -- --------- --- --- -- -- --

--------, --------, ---------, --- ---, --- ---, ---- ----, -- -- --, -- -- --, -- -- --.

## SOLID LINES AND DIVIDED LINES

Thus we see that a line is to be whole, divided into two, three or more parts. The ancient Chinese systems of divination, the I Ching and the T=ai Hsüan Ching, utilized lines and their divisions. The number of lines can be two, bigram, three, trigram, four, tetragram, or six, hexagram. The I Ching uses the hexagram, 6 lines each being whole or divided in two. The T=ai Hsüan Ching uses the tetragram, 4 lines each being whole, divided in two or divided in three. Each can be analyzed in terms of basic two set lines, that are then increased by reflection or mirroring. The I Ching recognized dual forces at work in the universe, the female Yin, and the male Yang. The T=ai Hsüan Ching recognizes three forces, the T=ien and the Ti balanced by the Jen. All forces are part of the T=ai Chi, the Great Unity.

## DUALITY OF VIEW

An important aspect of the lines is that they can either be considered as lines or spaces. Are we looking at the black or the white, at foreground or background? A solid line is 1 line that has 0 space. A split line has 2 parts or 1 space. A three part line has 3 parts or 2 spaces. Thus, we can consider the lines as 1, 2, and 3, or 0, 1, and 2.

This is not unlike the numbering of the Hebrew letter-numbers, the standard starting with 1 and the lost starting with 0. This is a clear example of duality.

## THE I CHING: POWERS OF 2

The I Ching divides each line into a whole or Yang, and in two parts, or Yin. Thus, the bottom or first line has two possibilities, whole, Yang or in two parts, Yin. A bigram has two lines, and the lower line can be either Yang or Yin, and each of these can have Yang or Yin for the second line. Thus, there are 2 x 2 = 4 possibilities for an I Ching bigram. If the whole line is labeled a 1, and the two part line, a 2, then the first line is either 1 or 2, and the bigrams are: 1 1; 1 2; 2 1; 2 2. 1 1 is called the Greater Yang, 2 2 is called the Greater Yin, 1 2 is called the Lesser Yang, and 2 1 is called the Lesser Yin. Adding a third line, we get a trigram, and each of the four bigrams can have a Yang or a Yin for the third line. This gives us 2 x 4 = 8 possibilities. Adding a fourth line, now the system is a tetragram, we get an additional factor of 2 for the number of possibilities. Thus, for an I Ching tetragram, 2 x 8 = 16 possibilities. Adding another line and we have 2 x 16 = 32 possibilities, and finally, for six lines, a hexagram, there are 2 x 32 = 84 possible different combinations.

Rule: to get the number of possibilities for an I Ching gram, raise the base of 2 to the number of lines, 2number of lines = number of possibilities. Thus, for the usual hexagram, or six lines, we get 26 = 64 grams.

The first and major line has three possibilities, whole or 1, in two parts or 2, and in three parts or 3. The bigram adds a second line that can be a 1, 2 or 3. Thus, a T=ai Hsüan Ching bigram can have 3 x 3 possibilities. For bigrams the number determination is as follows:

- possibility 1 has lower and upper line whole, 1 1
- possibility 2 has lower line whole and upper line divided in two, 1 2
- possibility 3 has lower line whole and upper line in three parts, 1 3
- possibility 4 has lower line in two parts and upper line whole, 2 1
- possibility 5 has lower line in two parts and upper line in two parts, 2 2
- possibility 6 has lower line in two parts and upper line in three parts, 2 3
- possibility 7 has lower line in three parts and upper line whole, 3 1
- possibility 8 has lower line in three parts and upper line in two parts, 3 2
- possibility 9 has lower line in three parts and upper line in three parts, 3 3.

We use the nomenclature of calling the lower line first, and the upper line second. If the line is complete we label it with a 1, if in two parts with a 2, and if in three parts with a 3. Thus, 2 3 means that the lower line is in two parts and that the upper line is in three parts. This would indicate that 2 3 refers to possibility 6.

If we go to a trigram or three lines, then the number of possibilities increase by a factor of 3, or 3 x 9 = 27 possibilities. For a tetragram, or four lines, the possibilities increase by

another factor of 3, or 3 x 27 = 81. For five lines, the possibilities are 3 x 81 = 243, and finally for a hexagram, six lines, 3 x 243 = 729. The general rule is given by 3 raised to the number of lines, or 3number of lines = number of possibilities.

## I CHING MAGIC SQUARES

We recall that a magic square is a 3 x 3, 4 x 4, etc arrangements of numbers. Thus, the first magic square would be a T'ai Hsüan Ching bigram having 9 possibilities, a 3 x 3 magic square. The next would be an I Ching tetragram having 16 possibilities, a 4 x 4 magic square. The next would be an I Ching hexagram having 64 possibilities, an 8 x 8 magic square. This is followed by a T=ai Hsüan Ching tetragram having 81 possibilities, a 9 x 9 magic square. If we restrict the maximum number of lines to 6, namely to a hexagram, the last is a T=ai Hsüan Ching hexagram having 729 possibilities, a 27 x 27 magic square.

Naturally, there are more magic squares, such as the I Ching 8 line that has 256 possibilities, or a

16 x 16 magic square. We shall call all of these the Ching Magic Squares.

## CHING 3 X 3 MAGIC SQUARE

The T=ai Hsüan Ching bigram has 9 possibilities, 32 = 9, and relates to a 3 x 3 magic square. We can use this system to replace the usual numbers in a 3 x 3 magic square.

| 4 | 9 | 2 |
|---|---|---|
| 3 | 5 | 7 |
| 8 | 1 | 6 |

by each of the possibilities. It is important that we use a consistent consecutive method of numbering the possibilities. We start with the lowest line and number consecutively, adding the same consective numbering to each of the succeeding lines. Thus, 1 = 1 1; 2 = 1 2; 3 = 1 3; 4 = 2 1; 5 = 2 2; 6 = 2 3; 7 = 3 1; 8 = 3 2; and 9 = 3 3. Our 3 x 3 magic square becomes:

| 2 1 | 3 3 | 1 2 |
|-----|-----|-----|
| 1 3 | 2 2 | 3 1 |
| 3 2 | 1 1 | 2 3 |

Notice the this is still a magic square, where the sum of every first digit adds to 6, as does the sum of every second digit. Thus, each row, column and diagonal, adds to 66. Thus, both original and line number specifications are magic. Symmetric cells add to 4 4.

## I CHING 4 X 4 EVENLY-EVEN MAGIC SQUARE

This magic square has 16 elements and is based on the 16 possibilities of the I Ching tetragrams. For this case, we have 24 = 16 possibilities. These can be numbered as follows: 1 = 1 1 1 1; 2 = 1 1 1 2;

3 = 1 1 2 1; 4 = 1 1 2 2; 5 = 1 2 1 1; 6 = 1 2 1 2; 7 = 1 2 2 1; 8 = 1 2 2 2; 9 = 2 1 1 1; 10 = 2 1 1 2; 11 = 2 1 2 1; 12 = 2 1 2 2; 13 = 2 2 1 1; 14 = 2 2 1 2; 15 = 2 2 2 1; and 16 = 2 2 2 2.

| 4 | 14 | 15 | 1 |
|---|----|----|---|
| 9 | 7 | 6 | 12 |
| 5 | 11 | 10 | 8 |
| 16 | 2 | 3 | 13 |

This becomes the 4 x 4 evenly-even Ching magic square:

| 1 1 2 2 | 2 2 1 2 | 2 2 2 1 | 1 1 1 1 |
|---------|---------|---------|---------|
| 2 1 1 1 | 1 2 2 1 | 1 2 1 2 | 2 1 2 2 |
| 1 2 1 1 | 2 1 2 1 | 2 1 1 2 | 1 2 2 2 |
| 2 2 2 2 | 1 1 1 2 | 1 1 2 1 | 2 2 1 1 |

This Ching magic square has unusual symmetry properties. Choosing any place, the digits of that place are a magic square, and each place has the magic sum of 6. Thus, all of the places add to 6 6 6 6. For example, if we choose the third place, the first row of third place numbers are: 2 1 2 1 which adds to 6. Furthermore, all cells that are symmetric, the digits of every place add to 3. For example, the first row, first column cell is a 1 1 2 2, and its symmetric cell is the last row, last column, which contains a 2 2 1 1. We note that 1 1 2 2 and 2 2 2 1 add to 3 3 3 3.

**I CHING 8 X 8 EVENLY-EVEN MAGIC SQUARE**

An 8 x 8 magic square has 64 cells, and is equal to 26 = 64 I Ching Hexagram.

EVEN ORDER: 8x8 (EVENLY-EVEN) STANDARD MAGIC SQUARE

| 8 | 58 | 62 | 5 | 4 | 59 | 63 | 1 |
|---|----|----|---|---|----|----|---|
| 49 | 15 | 51 | 12 | 13 | 54 | 10 | 56 |
| 17 | 42 | 22 | 44 | 45 | 19 | 47 | 24 |
| 32 | 39 | 35 | 29 | 28 | 38 | 34 | 25 |
| 40 | 31 | 27 | 37 | 36 | 30 | 26 | 33 |
| 41 | 18 | 46 | 20 | 21 | 43 | 23 | 48 |
| 9 | 55 | 11 | 52 | 53 | 14 | 50 | 16 |
| 64 | 2 | 6 | 61 | 60 | 3 | 7 | 57 |

The cells can be ordered as follows:

| 8 | 58 | 62 | 5 | 4 | 59 | 63 | 1 |
|---|----|----|---|---|----|----|---|
| 49 | 15 | 51 | 12 | 13 | 54 | 10 | 56 |
| 17 | 42 | 22 | 44 | 45 | 19 | 47 | 24 |
| 32 | 39 | 35 | 29 | 28 | 38 | 34 | 25 |
| 40 | 31 | 27 | 37 | 36 | 30 | 26 | 33 |
| 41 | 18 | 46 | 20 | 21 | 43 | 23 | 48 |
| 9 | 55 | 11 | 52 | 53 | 14 | 50 | 16 |
| 64 | 2 | 6 | 61 | 60 | 3 | 7 | 57 |

**I CHING 9 X 9 MAGIC SQUARE**

A 9 x 9 magic square contains 81 cells, and therefore 34 = 18 possibilities. It is therefore a T=ai Hsüan Ching tetragram. This would lead to a 9 x 9 square composed of tetragrams. Following the pattern of the double line, with lines having a one, two, and three part priority, the first Shou number would have 1 1 as the top two lines and 1 1 as the bottom two lines. This is followed by the second Shou number with 1 1 on top and 1 2 on the bottom. Shou number three, would have 1 1 on top and 1 3 on the bottom. Shou number four, has 1 1 on top and 2 1 on the bottom. Shou number 9 has 1 1 on top and 3 3 on the bottom. Shou 10 has 1 2 on top and 1 1 on bottom. Shou 11 has 1 2 on top and 1 2 on bottom. Shou 18 has 1 2 on top 3 3 on bottom. This pattern is continued to Shou 81 which has 3 3 on top and 3 3 on the bottom.

Recap: Each set of nine numbers is has the same upper top numbers. Shou 1 - 9 have top numbers of 1 1; 10 - 18 have top numbers of 1 2; 19 - 27 have 1 3; 28 - 36 have top numbers of 2 1; 37 - 45 have top numbers of 2 2; 46 - 54 have top numbers of 2 3; 55 - 63 have top numbers of 3 1; 64 - 72 have 3 2; and 73 - 81 have 3 3. A very ordered numbering system. We can order from the top or from the bottom, here we are ordering from the top. The ordering does not change the analysis or the symmetry of the resulting Ching magic square. The tetragrams numbers are therefore 1 1 1 1; 1 1 1 2; 1 1 1 3; 1 1 2 1; 1 1 2 2; 1 1 2 3; 1 1 3 1; 1 1 3 2; 1 1 3 3; 1 2 1 1; etc.

## ODD ORDER: 9X9 STANDARD MAGIC SQUARE

| 37 | 78 | 29 | 70 | 21 | 62 | 13 | 54 | 5 |
|----|----|----|----|----|----|----|----|----|
| 6 | 38 | 79 | 30 | 71 | 22 | 63 | 14 | 46 |
| 47 | 7 | 39 | 80 | 31 | 72 | 23 | 55 | 15 |
| 16 | 48 | 8 | 40 | 81 | 32 | 64 | 24 | 56 |
| 57 | 17 | 49 | 9 | 41 | 73 | 33 | 65 | 25 |
| 26 | 58 | 18 | 50 | 1 | 42 | 74 | 34 | 66 |
| 67 | 27 | 59 | 10 | 51 | 2 | 43 | 75 | 35 |
| 36 | 68 | 19 | 60 | 11 | 52 | 3 | 44 | 76 |
| 77 | 28 | 69 | 20 | 61 | 12 | 53 | 4 | 45 |

# I CHING 9 X 9 MAGIC SQUARE

| | | | | | | | | |
|---|---|---|---|---|---|---|---|---|
| 2 2 1 1 | 3 3 2 3 | 2 1 1 2 | 3 2 3 1 | 1 3 1 3 | 3 1 3 2 | 1 2 2 1 | 2 3 3 3 | 1 1 2 2 |
| 1 1 2 3 | 2 2 1 2 | 3 3 3 1 | 2 1 1 3 | 3 2 3 2 | 1 3 2 1 | 3 1 3 3 | 1 2 2 2 | 2 3 1 1 |
| 2 3 1 2 | 1 1 3 1 | 2 2 1 3 | 3 3 3 2 | 2 1 2 1 | 3 2 3 3 | 1 3 2 2 | 3 1 1 1 | 1 2 2 3 |
| 1 2 3 1 | 2 3 1 3 | 1 1 3 2 | 2 2 2 1 | 3 3 3 3 | 2 1 2 2 | 3 2 1 1 | 1 3 2 3 | 3 1 1 2 |
| 3 1 1 3 | 1 2 3 2 | 2 3 2 1 | 1 1 3 3 | 2 2 2 2 | 3 3 1 1 | 2 1 2 3 | 3 2 1 2 | 1 3 3 1 |
| 1 3 3 2 | 3 1 2 1 | 1 2 3 3 | 2 3 2 2 | 1 1 1 1 | 2 2 2 3 | 3 3 1 2 | 2 1 3 1 | 3 2 1 3 |
| 3 2 2 1 | 1 3 3 3 | 3 1 2 2 | 1 2 1 1 | 2 3 2 3 | 1 1 1 2 | 2 2 3 1 | 3 3 1 3 | 2 1 3 2 |
| 2 1 3 3 | 3 2 2 2 | 1 3 1 1 | 3 1 2 3 | 1 2 1 2 | 2 3 3 1 | 1 1 1 3 | 2 2 3 2 | 3 3 2 1 |
| 3 3 2 2 | 2 1 1 1 | 3 2 2 3 | 1 3 1 2 | 3 1 3 1 | 1 2 1 3 | 2 3 3 2 | 1 1 2 1 | 2 2 3 3 |

This square possesses the remarkable qualities that choosing the first, second, third, or fourth digit in every row, column and diagonal, that selected digit will always add to 18. Symmetric cells add to 4 4 4 4. While all magic cells add to 18 18 18 18.

## HEXAGRAMS AND 9 X 9 X 9

It is possible to form a magic cube that contains 9x9x9 = 729 cells or boxes. This number can be arrived at using hexagrams, 6 lines that can be whole, in two or in three parts. There are 729 different ways that the hexagram can be divided into one two and three parts. 729 = 9x9x9 = 3x3x3x3x3x3 = 36.

There are 64 ways that a hexagram can divided into one and two parts, and thus this is amenable to an 8 x 8 magic square.

# SERIES BIBLIOGRAPHY

**A**

Aberbach, Moses and Bernard Grossfeld trans. and analysis. Targum Onkelos to Genesis. Center for Judaic Studies. Denver, Colorado: KTAV Publishing House, Inc., 1982.

Aharoni, Yohanan and Michael Avi-Yonah. Bible Atlas. New York: Macmillan Publishing Company, 1977.

Andersen, Francis I. and Forbes, A. Dean. Biblical Hebrew Grammar Visualized. Winona Lake, Indiana, Eisenbrauns, 2012

Andrews, W.S. Magic Squares and Cubes. New York: Dover Publications, Inc., 1960.

Arberry, Arthur J. The Koran Interpreted. New York: Macmillan Publishing Company, 1955.

Arguelles, Jose. Earth Ascending. Colorado: Shambhala Publications, Inc., 1984.

Arguelles, Jose and Miriam. Mandala. Colorado: Shambhala Publications, Inc., 1972.

Ashlag, Rabbi Yehuda. An Entrance to the Tree of Life. Edited by Philip S. Berg. Israel: Research Centre for Kabbalah, 1977.

——— An Entrance to the Zohar. Edited by Philip S. Berg. Israel: Research Centre for Kabbalah, 1974.

——— The Kabbalah: A Study of the Ten Luminous Emanations from Rabbi Isaac Luria, Vol. I. Translated by Rabbi Levi I. Krakovsky. New York: Research Centre for Kabbalah, 1969.

**B**

Baer, Randall N. and Vicki V. The Crystal Connection. San Francisco: Harper and Row Publishers, 1986.

Baigent, Michael and Richard Leigh. The Dead Sea Scrolls Deception. New York: Simon and Schuster, 1991.

Baker, Dr. Douglas. The Opening of the Third Eye. Great Britain: The Aquarian Press, 1977.

Balin, Peter. The Flight of Feathered Serpent. Wisconsin: Arcana Publishing Company, 1978.

Barnsley, Michael. Fractals Everywhere. Boston: Academic Press, Inc., 1988.

Bauval, Robert and Adrian Gilbert. The Orion Mystery. New York: Crown Publishers, Inc., 1994.

ben Joseph, Rabbi Akiva. The Book of Formation: Sefer Yetzirah. New York: KTAV Publishing House, Inc., 1970.

ben Shimon Halevi, Z'ev. Kabbalah. London: Thames and Hudson Ltd., 1979.

Benson, William H. and Oswald Jacoby. New Recreations with Magic Squares. New York:Dover Publications Inc., 1976.

Berg, Philip S. Kabbalah for the Layman. New York: Research Centre for Kabbalah, 1981.

——— Power of Aleph Beth. 2 vols. New York: Research Centre for Kabbalah, 1988.

——— The Wheels of a Soul. Israel: Research Centre for Kabbalah, 1984.

Blavatsky, H.P. The Key to Theosophy. California: Theosophical University Press, 1987.

——— The Secret Doctrine. Vol I, Cosmogenesis. California: Theosophical University Press, 1977.

——— The Secret Doctrine. Vol. II, Anthropogenesis. California: Theosophical University Press, 1977.

——— Isis Unveiled. Vols. I, II, Science. California: Theosophical University Press, 1976.

Bloch, Abraham P. The Biblical and Historical Background of the Jewish Holy Days. New York: KTAV Publishing House, Inc., 1978.

Blumenthal, David R. Understanding Jewish Mysticism. Vol. I, The Merkabah Tradition and the Zoharic Tradition. New York: KTAV Publishing House, Inc., 1978.

——— Understanding Jewish Mysticism. Vol. II, The Philosophical - Mystical Tradition and the Hasidic Tradition. New York: KTAV Publishing House, Inc., 1982.

Bokser, Rabbi Ben Zion. From the World of the Cabbalah. New York: Philosophical Library, 1954.

——— The Wisdom of the Talmud. New York: Philosophical Library, 1951.

Breasted, James Henry. Ancient Records of Egypt. 3 vols. New York: Russell and Russell, Inc., 1906.

Brown, Francis, ed. A Hebrew and English Lexicon of the Old Testament. Oxford: Clarendon
        Press.
Brown, Peter Lancaster. Megaliths, Myths, and Men. New York: Harper Colophon Books, 1976.
Brunstein, Karl. Beyond the Four Dimensions. New York: Walker and Company, 1979.
Buckingham, Jamie. Power for Living. United States: Arthur S. De Moss Foundation, 1983.
Buckland, Raymond. Practical Color Magick. Minnesota: LLewellyn Publications, 1983.
Buddhist Bible, A. Edited by Goddard, Dwight. Boston: Beacon Press, 1970.
Budge, Wallis E.A. The Book of the Dead. New York: Dover Publications, Inc., 1967.
—— The Gods of the Egyptians. 2 vols. New York: Dover Publications, Inc., 1969.
—— Egyptian Language. New York: Dover Publications, Inc., 1978.


C
Case, Paul Foster. The Book of Tokens. Los Angeles: Builders of the Adytum, 1968.
Chase, Mary Ellen. Life and Language in the Old Testament. New York: W.W. Norton and
Company, Inc., 1955.
Clagett, Marshall. Greek Science in Antiquity. New York: Collier Books, 1955.
Chu, W.K., trans. The Astrology of I Ching. London: Routledge and Kegan Paul, 1980.
Coe, Michael D. and Stone, Mark Van. Reading the Maya Glyphs. London: Thames and Hudson,
        2005.
Comay, Joan. Who's Who in the Old Testament. New York: Holt, Rinehart and Winston, 1971.
Crowley, Aleister. 777 and Other Qabalistic Writings. Maine: Samuel Weiser, Inc., 1986.
Culi, Rabbi Yaakov. The Torah Anthology: MeAm Lo'ez. Vols. 1-4. Translated by Rabbi Aryeh
Kaplan. New York: Maznaim Publishing Corporation, 1977.


D
Davis, Philip. The Lore of Large Numbers. New York: Random House, 1961.
Davis, Philip and Reuben Hersh. The Mathematical Experience. Boston: Houghton
        Miflin Company, 1981.
Davies, Paul Ed. The New Physics. Great Britain: Cambridge University Press, 1993.
de Nicolas, Antonio T. Avatara. New York: Nicolas Hays, Ltd., 1976.
—— Four-Dimensional Man. New York: Nicolas Hays, Ltd., 1976.
de Rola, Stanislas Klossowski. Alchemy. London: Thames and Hudson Ltd., 1973.
Dobin, Rabbi Joel C. To Rule Both Day and Night. New York: Inner Traditions International, 1977.
d'Olivet, Fabre. The Hebraic Tongue Restored. Translated by Nayan Louise Redfield. New York:
Samuel Weiser, Inc., 1978.
Drury, Nevill and Andrew Watson. Healing Music. New York: Avery Publishing Group, Inc., 1987.


E
Einstein, Albert. Relativity, The Special and the General Theory. Authorized Translation by Robert
W. Lawson. New York: Crown Publishers, Inc. 1961.
Eisen, William. The English Caballah. Vol. I, The Mysteries of Pi. California: DeVorss and
Company, 1980.
—— The English Caballah. Vol. II, The Mysteries of Phi. California: DeVorss and Company, 1982.
—— The Universal Language of Cabbalah. The Master Key of the God Consciousnous.
        California: DeVorss and Company, 1989.
Eisenberg, Azriel. The Book of Books. London: The Soncino Press. 1976.
Eisenman, Robert and Michael Wise. The Dead Sea Scrolls Uncovered. New York: Barnes and
        Noble Inc., 1994.
Epstein, Perle. Kabbalah the Way of the Jewish Mystic. New York: Doubleday and Company, Inc.,
        1978.

Evans-Wentz, W.Y. ed. The Tebetan Book of the Dead. New York: Oxford Press, 1960.


F
Feldman, W.M. Rabbinical Mathematics and Astronomy. New York: Sefer-Hermon Press,Inc.,1978.
Fell, Barry. America B.C. New York: Pocket Books, 1989.
Fine, Lawrence, ed. Safed Spirituality. New Jersey: Paulist Press, 1984.
Friedman, A.Z. Wellsprings of Torah: An Anthology of Biblical Commentaries. 2 vols. New York: The Judaica Press, Inc., 1969.
Fuller, J.F.C. The Secret Wisdom of the Qabalah. London: Rider & Co.


G
Gaster, Theodor H. The Dead Sea Scriptures. New York: Anchor Books, 1976.
—- Myth, Legend, and Custom in the Old Testament. United States: Harper and Row, Publishers, 1969.
Ginzberg, Louis. Legends of the Bible. Philadelphia: The Jewish Publication Society of America, 1968.
Glazerson, M. Sparks of the Holy Tongue. New York: Feldheim Publishers, 1981.
Godwin, Joscelyn. The Mystery of the Seven Vowels. Michigan: Phanes Press, 1991.
Graves, Robert. The Greek Myths. England: Penguin Books, 1992.
Graetz, Heinreich. History of the Jews. 6 vols. Philadelphia: The Jewish Publication Society of America, 1891.
Gruberger, Philip S. The Kabbalah: A Study of the Ten Luminous Emanations. Vol. II, Circles and Straightness. New York: Research Centre for Kabbalah, 1973.


H
Hall, Donald E. Musical Acoustics. California: Wadsworth Publishing Company, 1980.
Hall, Manley, P. The Secret Teachings of All Ages. California: The Philosophical Research:Society, Inc., 1977.
The Holy Scriptures. Philadelphia: The Jewish Publication Society of America, 1955.
Hapgood, Charles H. Maps of the Ancient Sea Kings. Philadelphia: Chilton Books, 1966.
Hardy, G. H. and Wright, E. M. An Introduction to the Theory of Numbers. Great Britain: Clarendon Press, Oxford, 1990.
Hawking, Stepehn W. A Brief History of Time, From the Big Bang to Black Holes, Toronto: Bantam Books, 1988.
Huntley, H.E. The Divine Proportion. New York: Dover Publications, Inc., 1970.
Hurtak, J.J. The Book of Knowledge: The Keys of Enoch. California: The Academy for Future Science, 1977.


I
Idel, Moshe. Kabbalah. Connecticut: Yale University Press, 1988.
—- Language, Torah, and Hermeneutics in Abraham Abulafia. New York: State University of New York Press, 1989.
Ifrah, Georges. From One to Zero. Translated by Lowell Blair. New York: Penquin Books, 1985.
Itten, Johannes. The Elements of Color. Translated by Ernst Van Hagen. New York: Van Nostrand Reinhold Company, 1970.


J
The Jerusalem Bible. New York: Doubleday and Company, Inc., 1968.
Josephus. Kregel Publications: Grand Rapids Michigan, 1977.

K

Kalisch, Isadore. The Sepher Yetzirah. New York: L. H. Frank & Co., 1877.

Kals, W.S. Stars and Planets. San Francisco: Sierra Club Books, 1990.

Kaplan, Aryeh, Trans. The Bahir. New York: Samuel Weiser, Inc., 1979.

—- Jewish Meditation. New York: Schocken Books, Inc., 1985.

—- The Light Beyond. New York: Maznaim Publishing Corporation, 1981.

—- The Living Torah. New York: Maznaim Publishing Corporation, 1981.

—- Meditation and the Bible. Maine: Samuel Weiser, Inc., 1978.

—- Meditation and Kabbalah. Maine: Samuel Weiser, Inc., 1982.

—- Sefer Yetzirah. Maine: Samuel Weiser, Inc., 1990.

—- Waters of Eden. New York: National Conference of Synagogue Youth/Union of Orthodox
Jewish Congregations of America, 1982.

Khanna, Madhu. Yantra. London: Thames and London Ltd., 1979.

King, C.W. The Gnostics and Their Remains. Secret Doctrine Reference Series. San Diego: Wizards
          Bookshelf, 1982.

Kitov, Eliyahu. The Book of our Heritage. 3 vols. Translated from the Sefer Hatoda'ah by Nathan
Bulman. New York: Feldheim Publishers, 1973.

Kolatch, Alfred J. This is the Torah. New York: Jonathan David Publishers, Inc., 1988.

Krakovsky, Rabbi Levi Isaac. Kabbalah: The Light of Redemption. Israel: Press of the "Yeshivat
Kol Yehuda," 1970.


L

Lauf, Detlef Ingo. Secret Doctrines of the Tibetan Books of the Dead. Boston, Massachusetts:
          Shambhala Publications, Inc., 1977.

Laurence, Richard. Trans. The Book of Enoch the Prophet. Minneapolis: Wizards Bookshelf, 1976.

Lawlor, Robert and Deborah. Trans. Mathematics Useful for Understanding Plato. Secret Doctrine
Reference Series. San Diego: Wizards Bookshelf, 1979.

—- Theon of Smyrna. Secret Doctrine Reference Series. San Diego: Wizards Bookshelf, 1979.

Laymon, Charles M., Editor. The Interpreter=s One-Volume Commentary on the Bible. Nashville
          and New York: Abingdon Press, 1971.

Leadbeater, C.W. Ancient Mystic Rites. Illinois: The Theosophical Publishing House, 1986.

—- The Masters and the Path. Illinois: The Theosophical Publishing House, 1925.

Leaf, Reuben. Hebrew Alphabets. United States: Bloch Publishing Company, Inc., 1976.

LePlongeon, Augustus. Sacred Mysteries Among the Mayas and the Quiches. San Diego: Wizards
          Bookshelf, 1985.

Levi, Eliphas. The Book of Splendours. Great Britain: The Aquarian Press and Samuel Weiser, Inc.,
          1981.

Lipman, Eugene J. The Mishnah. New York: The Viking Press, 1970.

Locks, Gurman G. The Spice of the Torah - Gematria. New York: Judaica Press, 1985.

Luzzatto, Rabbi Moses. The General Principles of Kabbalah. Israel: Research Centre of Kabbalah,
          1984.


M

Maimonides, Moses. The Guide of the Perplexed. Vol I and II. Translated by Shlomo Pines.
          Chicago: University of Chicago Press, 1963.

Malbim, Rabbi Meir. Commentary on the Torah. Israel: Hillel Press, 1978.

Martin, Bernard. A History of Judaism. 2 vols. New York: Basic Books, Inc., 1974.

Mathers, S.L. Magregor. The Kabbalah Unveiled. York Beach, Maine: Samuel Weiser, Inc.

Mazar, Benjamin, and Davis, Moshe. The Illustrated History of the Jews. Israeli Publishing Institute,
          1963.

McClain, Ernest G. Meditations through the Quran. York Beach, Maine: Nicolas Hays, Inc., 1981.
—– The Myth of Invariance. New York: Nicolas Hays, Ltd., 1976.
—– The Pythagorean Plato: Prelude to the Song Itself. New York: Nicolas Hays, Ltd., 1978.
Montgomery, John.
—–Dictionary of Maya Hieroglyphs. New York: Hippocrene Books, Inc., 2002.
—– How to Read Maya Hieroglyphs. New York: Hippocrene Books, Inc., 2002.
Mookerjee, Ajit. Kundalini- The Arousal of the Inner Enerby. New York: Destiny Books, 1982.
Moravesik, Michael J. Musical Sound. New York: Paragon House Publishers, 1987.
Mordell, Phineas. The Origin of Letters and Numerals according to the Sefer Yetzirah. New York:
        Samuel Weiser Inc., 1975.
Morley, Sylvanus G. The Ancient Maya. Stanford, California: Stanford University Press. 1956.
Morrisio, Wolff and Fraknoi. Abell's Exploration of the Universe. Orlando, Florida: Saunders
College Publishing, 1995.
Munk, Rabbi Michael J. The Wisdom of the Hebrew Alphabet. ArtScroll Mesorah Series. New
        York: Mesorah Publications, Ltd., 1983.
Musashi, Myamoto. A Book of Five Rings. Translated by Victor Harris. Woodstock, New York:
        The Overlook Press, 1974.
Myer, Isaac. Quabbalah. California: Wizards Bookshelf, 1988.


N
Nelson, Ralph, trans. Popol Vuh. Boston: Houghton Miffin Company, 1976.
Newton, Isaac. The Chronology of Ancient Kingdoms Amended. Middlesex: The Echo Library,
        2007.
Norelli-Cachelet, Partizia. The Gnostic Circle. New York: Samuel Weiser, Inc., 1975.
Norvill, Roy. Hermes Unveiled. Great Britain: Ashgrove Press Limited, 1986.


O
Oliver, George. The Pythagorean Triangle. Secret Doctrine Reference Series. San Diego: Wizards
        Bookshelf, 1984.
Orlinsky, Harry M. Understansing the Bible Through History and Archeology. New York:KTAV
        Publishing House, Inc., 1972.


P
Papus. The Qabalah. Great Britain: The Aquarian Press, 1977.
—– The Tarot of the Bohemians. New York: Samuel Weiser, Inc., 1958.
Payne, Richard J. ed. Zohar. Translated by Matt, Daniel Chanen. New Jersey: Paulist Press, 1983.
Pedretti, Carlo. Leonardo Da Vinci. TAJ Books, 2004.
Pennick, Nigel. Magical Alpahbets. Maine: Samuel Weiser, Inc., 1992.
The Editors of Pensee. Velikovsky Reconsidered. New York: Doubleday and Company, Inc., 1976.
The Pentateuch and Rashi's Commentary. 5 vols. Translated by Sharfman, Rabbi Benjamin and
Rabbi Abraham Ben Isaiah. New York: S.S. and R. Publishing Company, Inc., 1949.
Podwal, Mark. A Book of Hebrew Letters. Philadelphia: The Jewish Publication Society of
        America, 1978.
Politoske, Daniel T. Music. Englewood Cliffs, New Jersey: Prentice-Hall, Inc., 1974.

Ponce, Charles. Kabbalah. United States: Quest Books, 1973.
Pope, Marvin. Song of Songs. The Anchor Bible, Vol 7c. Garden City, New York: Double Day and
        Company, Inc., 1977.
Popol Vuh. The Great Mythological Book of the Ancient Maya,Translated by Ralph Nelson.
Boston: Houghton Mifflin Company,1974.

Prophet, Elizabeth Clare. Forbidden Mysteries of Enoch. United States: Summit University Press, 1983.

R

Rameau, Jean-Philippe. Treatise on Harmony. New York: Dover Publications, Inc., 1971

Reade, Julian. Assyrian Sculpture. Cambridge, Massachusetts: Harvard University Press, 1983.

Regardie, Israel. Foundations of Practical Magic. Great Britain: The Aquarian Press Limited, 1979.

The Reubeni Foundation. Chronicles, New of the Past. 2 vol. Israel: The Arrow Company, 1970.

Room, Adrian. The Guiness Book of Numbers. Great Britain: Guiness Publishing Ltd., 1989.

Rosen, Dov. Shema Yisrael. Translated by Oachry, Leonard. 2 Vols. Israel: Peli Printing Works Ltd., 1972.

Rossing, Thomas D. The Science of Sound. New York: Addison-Wesley Publishing Company, 1990.

S

Sadhu, Mouni. Concentration. California: Wilshire Book Company, 1959.

—— In Days of Great Peace. California: Wilshire Book Company, 1957.

—— Meditation. California: Wilshire Book Company, 1967.

—— The Tarot. California: Wilshire Book Company, 1968.

—— Ways to Self-Realization. California: Wilshire Book Company, 1962.

Sargeant, Winthrop, trans. The Bhagavad Gita. Albany: State University of New York Press, 1984.

Sarna, Nahum M. Understanding Genesis. New York: Schocken Books, 1966.

Sayce, A.H. Astronomy and Astrology of the Babylonians. San Diego, California: Wizards Bookshelf, 1981.

Schroeder, Gerald L. Genesis and the Big Bang. New York: Bantam Books, 1990.

Scnimmel, Annemarie. The Mystery of Numbers. New York: Oxford Press, 1993.

Scholem, Gershom G. Zohar. New York: Schocken Books Inc., 1949.

—— Kabbalah. New York: New American Library, 1974.

—— Jewish Gnosticism, Merkabah, Mysticism, and Talmudic Tradition. New York: The Jewish Theological Seminary of America, 1965.

—— Major Trends in Jewish Mysticism. New York: Schocken Books Inc., 1971.

—— On the Kabbalah and its Symbolism. United States: Schocken Books Inc., 1965.

Scherman, Rabbi Nosson and Meir Zlorowitz, eds. The Book of Ezekiel. ArtScroll Tanach Series. New York: Mesorah Publications, Ltd., 1977.

—— Genesis. ArtScroll Tanach Series. New York: Mesorah Publications, Ltd., 1977.

—— Song of Songs. ArtScroll Tanach Series. New York: Mesorah Publications, Ltd., 1977.

Shahn, Ben. The Alphabet of Creation. New York: Schocken Books, 1954.

The Pentatuech and Rashi's Commentary. Translated by Sharfman, Rabbi Benjamin and Rabbi Abraham Ben Isaiah. 5 Vols. New York: S.S. and R. Publications, Inc., 1949.

Shanks, Hershel, et al. The Dead Sea Scrolls after Forty Years. Biblical Archeology Society. Washington, D.C., 1990

Sharkey, John. Celtic Mysteries. Great Britain: Thames and Hudson Ltd., 1975.

Silver, Abba Hillel. Moses and the Original Torah. New York: The Macmillan Company, 1961.

Simon, Maurice and Harry Sperling, Translation. The Zohar. New York: The Soncino Press, 1984.

Sitchin, Zecharia. The 12th Planet. New York: Avon Books, 1976.

—— The Stairway to Heaven. New York: Avon Books, 1983.

—— The Wars of Gods and Men. New York: Avon Books, 1985.

—— The Lost Realms, New York: HarperCollins, 1990.

—— When Time Began, New York: HarperCollins, 1993.

—— The Cosmic Code, New York: HarperCollins, 1998.

—— The End of Days, New York: HarperCollins, 2007.

—— The Lost Book of Enki, Vermont: Bear & Company, 2002.

—— The Anunnaki Chronicles, Vermont: Bear & Company, 2015.

Skinner, J. Ralston. The Source of Measures. San Diego, California: Wizards Bookshelf, 1982.

Smith, George. The Chaldean Account of Genesis. Secret Doctrine Reference Series. Minneapolis, Minnesota: Wizards Bookshelf, 1977.

Steinsaltz, Rabbi Adin. The Talmud: The Steinsaltz Edition. New York: Random House, 1989.

Stieglitz, Robert R. Numerical Structuralism and Cosmogony in the Ancient Near East. J. Social Biol. Struct. 1982 5, 255-266.

Stone, Merlin. When God was a Woman. New York and London: Harcourt Brace Jovanovic, 1978.

Suares, Carlo. The Cipher of Genesis. Colorado: Shambhala Publications, Inc., 1967.

—— The Sefer Yetzira. Colorado: Shambhala Publications, Inc., 1968.


T

Tame, David. The Secret Power of Music. Rochester, Vermont: Desitny Books, 1984.

Taylor, Thomas, Trans. The Cratylus, Phaedo, Parmenides, Timaeus, and Critias of Plato. Secret Doctrine Series. Minneapolis, Minnesota: Wizards Bookshelf, 1976.

—— The Eleusinian and Bacchic Mysteries. Secret Doctrine Series. San Diego, California: Wizards Bookshelf, 1987.

—— The Theoretic Arithmetic of the Pythagoreans. York Beach, Maine: Samuel Weiser, Inc., 1983.

Temple, Robert K.G. The Sirius Mystery. Rochester, Vermont: Destiny Books, 1987.

Thorsson, Edred. Futhark - A Handbook of Rune Magic. York Beach, Maine: Samuel Weiser, Inc., 1984.

Three Initiates. The Kybalion. Chicago, Illinois: The Yogi Publication Society, 1940.

Tompkins, Peter. Secrets of the Great Pyramid. New York: Harper and Row, 1978.

Trachtenberg, Joshua. Jewish Magic and Superstition. New York: Behrman House,1939.

Tyson, Donald. Rune Magic. Minnesota: Llewellyn Publications, 1988.


V

Velikovsky, Immanuel. Worlds in Collision. New York: Dell Publishing Company, Inc., 1972.

—— Ages in Chaos. New York: Doubleday and Company, Inc., 1952.

—— Peoples of the Sea. New York: Doubleday and Company, Inc., 1977.

—— Ramses II and His Time: New York: Doubleday and Company, Inc., 1978.

Vivekananda, Swami. The Complete Works, Nine Volumes. India: Advaita Ashrama, 1997, 2006.


W

Walker, Barbara. The I Ching of the Goddess. San Francisco: Harper and Row, 1986.

Walters, Derek. The T'ai Hsuan Ching. Great Britain: The Aquarian Press, 1983.

Wang, Robert. The Qabalistic Tarot. York Beach, Maine: Samuel Weiser, Inc., 1983.

West, John Anthony. Serpent in the Sky. New York: McJulian Press, Inc., 1987.

Westcott, W. Wynn. Sefer Yetzirah. New York: Samuel Weiser, 1980.

Whiston, William, Trans. Josephus. Grand Rapids, Michigan: Kregel Publications, 196

Wilder, Alexander. New Platonism and Alchemy. Secret Doctrine Reference Series. Minneapolis, Minnesota: Wizards Bookshelf, 1975

# ADDENDUM
## Ancient Measurements of Space

In his book The Greek Myths, Robert Graves refers to the Castration of Uranus, who fathered the Titans on Mother Earth, after he had thrown his rebellious sons the Cyclops into Tartarus, a gloomy place in the Underworld, which lies as far distant from the earth as the earth does from the sky; it would take a falling anvil nine days to reach its bottom. Let us use this concept of an anvil falling in earth's gravity to calculate how far it would fall (if it remains under these conditions) in nine days.

This is an example of Uniform Accelerated Motion. We assume the simplest of conditions; zero initial velocity (falling), no air resistance (an anvil is heavy and solid), and no limiting distance. Special relativity effects are not considered in this formulation.

Here $s$ = distance (m),
$t$ = time (s),
$g = 9.8$ m/s$^2$ the acceleration due to earth' gravity.
Thus, $s = (1/2) gt^2$ since the initial velocity is zero.

Distance anvil falls in nine days.
$t = 9 \times 24 \times 3600$ s $= 777,600$ s
$s = 4.9 t^2 = 2.962842624 \times 10^{12}$ m.
The mean distance of Uranus from Earth is $2.72395 \times 10^{12}$ m. This is a remarkable technique achievement described in the ancient Greek Myth.
Let us see how this technique applies to all the planet distances from earth.

**All distances are in $10^6$ km**

| Planet | Min | Max | Days falling | Distance |
|---|---|---|---|---|
| Venus | 38.2 | 261 | 1 | 36.57 |
| Mars | 54.5 | 401.4 | 2 | 146.3 |
| Mercury | 77.3 | 221.9 | 3 | 329.2 |
| Jupiter | 588.5 | 968.5 | 4 | 585.2 |
| Saturn | 1205.5 | 1658.6 | 6 | 1317 |
| Uranus | 2580.6 | 3153.5 | 9 | 2963 |
| Pluto | 4284.7 | 7528 | 12 | 5267 |
| Neptune | 4319 | 4711 | 11 | 4426 |

**Remarkable results with the exception of Mercury.**

**To calculate the acceleration required to cover a distance of one light year in one year of falling**

a = 2c/t = 2x3x108/(365.25x24x3600) = 19 m/s²

Jupiter has its acceleration due to gravity of 23.6 m/s² which exceeds the previous requirement.

This concludes the So You Think You know trilogy.

✳ ✳ ✳ ✳ ✳ ✳ ✳ ✳ ✳ ✳ ✳ ✳ ✳ ✳ ✳ ✳ ✳ ✳ עולם בורא לאל שבח ונשלם תם ✳ ✳ ✳ ✳ ✳ ✳ ✳ ✳ ✳ ✳ ✳ ✳ ✳ ✳ ✳ ✳ ✳ ✳ ✳

# Author Remarks

This Trilogy is about ancient secrets that are hiding in plain sight. They exist in the expanded associations between letters and numbers. It took me over ten years to solve the most simple yet profound passage in the Sefer Yetzirah, a Kabbalistic document that was presumably given to Abraham to use to return the Creator to his throne. It contains a key passage involving the Hebrew Letters and their number associations called Gematria.. The passage goes as follows: …

'Ten Sefirot of Nothingness
     ten and not nine
     ten and not eleven
Understand with Wisdom
Be wise with Understanding
     Examine with them
     and probe from them
Make [each] thing stand on its essence
And make the Creator sit on his base."

Sefer Yetzirah
The Book of Creation
Aryeh Kaplan

Sefirot ≡ The ten emanations, or powers, in which the Creator reveals himself and creates the physical world.

Therefore this means that the first letter is not 1 but 0. 0 through 9 is 10, the correct answer. This looks like 9 but is 10.
Starting with o and going to 10 looks like 10 but is 11.

The correct gematria also requires that even numbers are female, odd numbers are male The smaller the number the greater its potency. Thus the greatest potency is 0, with all its female characteristics.

This trilogy is just the beginning of the journey unifying science.and theology. I hope it will be well studied in the future.

\* \* \* \* \* \* \* \* \* \* \* \* \* \* \* \* \* \* תם ונשלם שבח לאל בורא עולם \* \* \* \* \* \* \* \* \* \* \* \* \* \* \* \* \* \*

www.ingramcontent.com/pod-product-compliance
Lightning Source LLC
Chambersburg PA
CBHW080802180526
45168CB00006B/2297